Forschungsberichte aus dem Lehrstuhl für Regelungssysteme

Technische Universität Kaiserslautern

Band 12

Forschungsberichte aus dem Lehrstuhl

für Regelungssysteme

Technische Universität Kaiserslautern

Band 12

Herausgeber:

Prof. Dr. Steven Liu

Sanad Al-Areqi

Investigation on Robust Codesign Methods for Networked Control Systems

Logos Verlag Berlin

λογος

Forschungsberichte aus dem Lehrstuhl für Regelungssysteme
Technische Universität Kaiserslautern

Herausgegeben von
Univ.-Prof. Dr.-Ing. Steven Liu
Lehrstuhl für Regelungssysteme
Technische Universität Kaiserslautern
Erwin-Schrödinger-Str. 12/332
D-67663 Kaiserslautern
E-Mail: sliu@eit.uni-kl.de

Bibliografische Information der Deutschen Nationalbibliothek

Die Deutsche Nationalbibliothek verzeichnet diese Publikation in der Deutschen Nationalbibliografie; detaillierte bibliografische Daten sind im Internet über http://dnb.d-nb.de abrufbar.

ISBN 978-3-8325-4170-5
ISSN 2190-7897

Logos Verlag Berlin GmbH
Comeniushof, Gubener Str. 47,
10243 Berlin
Tel.: +49 (0)30 / 42 85 10 90
Fax: +49 (0)30 / 42 85 10 92
http://www.logos-verlag.de

Investigation on Robust Codesign Methods for Networked Control Systems

Untersuchung zu robusten Codesign Methoden für vernetzte Regelungssysteme

Vom Fachbereich Elektrotechnik und Informationstechnik

der Technischen Universität Kaiserslautern

zur Verleihung des akademischen Grades

Doktor der Ingenieurwissenschaften (Dr.-Ing.)

genehmigte Dissertation

von

M. Sc. Sanad Al-Areqi

geboren in Taiz/Jemen

D 386

Tag der mündlichen Prüfung:	22.10.2015
Dekan des Fachbereichs:	Prof. Dr.-Ing. Hans D. Schotten
Vorsitzender der Prüfungskommission:	Prof. Dipl.-Ing. Dr. Gerhard Fohler
1. Berichterstatter:	Prof. Dr.-Ing. Steven Liu
2. Berichterstatter:	Prof. Dr. Ir. W.P.M.H. (Maurice) Heemels

Contents

Preface

Motivation

Networked control systems (NCSs), where multiple control loops are closed over a common communication network, have received much attention in recent years due to their scientific interest and technological significance both in academia and industry. This is witnessed by several published surveys, special issues, and monographs as well as by numerous applications ranging from transportation (automobiles, trains, aircrafts, etc.) over industrial applications (manufacturing and process control) to infrastructure systems (power systems, water distribution systems, etc.).

In NCSs, communication and control interact very tightly with each other. For ensuring stability and/or optimal performance of such systems, implementation-aware design techniques are strongly required. This requirement stems from the fact that the occurrence of network-induced imperfections such as latencies and congestion might degrade the control performance or even lead to instability. Much research in recent years has therefore focused on control and scheduling codesign for networked control systems. By jointly designing the control algorithm and the network scheduling algorithm, network-induced imperfections can be avoided or influenced at the benefit of satisfying certain control requirements. Several codesign concepts have been proposed which can roughly be classified into *time-triggered* codesign and *event-triggered* codesign.

Within the time-triggered codesign framework, scheduling decision of resources to tasks and control update instants are coordinated by a global or local clock. Various time-triggered codesign methods using static or dynamic scheduling strategies have been proposed. What usually missing in these methods, however, is the explicit consideration of other network-induced imperfections such as variable transmission time, variable sampling instants, packet dropouts, or quantization effects. In other words, a little attention has been paid to the robustness of NCSs against network-induced imperfections within the time-triggered codesign schemes.

Within the event-triggered codesign framework, scheduling decision and control update instants are coordinated by an event generator rather than a clock. In contrast to time-triggered codesign, a little work has been done in event-triggered codesign and only for specific scenarios. It is thus fair to say that event-triggered control and scheduling codesign is in its infancy.

To sum up, two directions of research are necessary to fill the gap outlined above:

- Unifying and generalizing existing codesign methods w.r.t. robustness
- Providing novel robust control and scheduling codesign methods

The aim of this thesis is to contribute to both research directions.

Simulation Platform

Within the thesis, all computations and simulations have been performed on a PC with Intel® Core™i7 CPU 2.80 GHz processor, 8 GB RAM, and Microsoft® Windows 7 Version 6.1 (Build 7601: Service Pack 1, 64 bit) operating system using MATLAB®Version 7.13.0.564 (R2011b,win-64) with the Control System Toolbox™Version 9.2 (R2011b). Moreover, the YALMIP Version 02-Oct-2013 toolbox [Löf04] together with the MOSEK 7 solver [AA14] have been used for solving LMI optimization problems.

Acknowledgements

The results printed in this thesis would not exist, after the blessings of Allah, without the great supervision and contribution of many lovely people. First and foremost, I would like to express my warm thanks and incredible gratitude to my supervisor Prof. Dr.-Ing. Steven Liu for his unlimited support and encouragement during my research journey at the Institute of Control Systems in the Department of Electrical and Computer Engineering at the University of Kaiserslautern. Furthermore, I am incredibly grateful for the learning experience, the scientific freedom, and the opportunities he has always provided me with. I would also like to thank Prof. Dr. Ir. W.P.M.H. (Maurice) Heemels for his interest in my thesis and for joining the PhD committee as a reviewer.

Special thanks go to my (former) work colleagues Swen Becker, Markus Bell, Felix Berkel, Sebastian Caba, Filipe Figueiredo, Daniel Görges, Yanhao He, Thomas Janz, Fabian Kennel, Ali Kheirkhah, Jutta Lenhardt, Markus Lepper, Xiaohai Lin, Peter Müller, Tim Nagel, Sven Reimann, Stefan Simon, Christian Tuttas, Yun Wan, Hengyi Wang, Jianfei Wang, Wei Wu, and Yakun Zhou. They have been as a big family to me.

Finally, warm thanks from the deep of my heart to my parents and family for all the love and support over all the years. Most of all, I would like to thank my lovely wife Gehan for all the patience, care and happiness she has given me. This thesis is dedicated to her.

Kaiserslautern, April 2015

Sanad Al-Areqi

Notation

Throughout the thesis, scalars are denoted by lower- and upper-case non-bold letters ($a, b, \ldots, A, B, \ldots$), vectors by lower-case bold letters ($\boldsymbol{a}, \boldsymbol{b}, \ldots$), matrices by upper-case bold letters ($\boldsymbol{A}, \boldsymbol{B}, \ldots$), and sets by upper-case double-struck letters ($\mathbb{A}, \mathbb{B}, \ldots$).

Sets

$\mathbb{N}$	Set of positive integers
$\mathbb{N}_0$	Set of non-negative integers
$\mathbb{R}$	Set of real numbers
$\mathbb{R}_0^+$	Set of non-negative real numbers
$\mathbb{R}^+$	Set of positive real numbers

Operators

$\boldsymbol{A}^{-1}$	Inverse of matrix $\boldsymbol{A}$
$\boldsymbol{A}^T$	Transpose of matrix $\boldsymbol{A}$
$\boldsymbol{A} \succ \boldsymbol{0}$	Matrix $\boldsymbol{A}$ positive definite, i.e. $\boldsymbol{x}^T\boldsymbol{A}\boldsymbol{x} > 0 \ \forall \boldsymbol{x} \neq \boldsymbol{0}$
$\boldsymbol{A} \succeq \boldsymbol{0}$	Matrix $\boldsymbol{A}$ positive semidefinite, i.e. $\boldsymbol{x}^T\boldsymbol{A}\boldsymbol{x} \geq 0 \ \forall \boldsymbol{x}$
$\boldsymbol{A} \prec \boldsymbol{0}$	Matrix $\boldsymbol{A}$ negative definite, i.e. $\boldsymbol{x}^T\boldsymbol{A}\boldsymbol{x} < 0 \ \forall \boldsymbol{x} \neq \boldsymbol{0}$
$\boldsymbol{A} \preceq \boldsymbol{0}$	Matrix $\boldsymbol{A}$ negative semidefinite, i.e. $\boldsymbol{x}^T\boldsymbol{A}\boldsymbol{x} \leq 0 \ \forall \boldsymbol{x}$
$\mathrm{tr}(\boldsymbol{A})$	Trace of matrix $\boldsymbol{A}$
$\det(\boldsymbol{A})$	Determinant of matrix $\boldsymbol{A}$
$\lambda_{\min}(\boldsymbol{A})$	Minimum eigenvalue of matrix $\boldsymbol{A}$
$\lambda_{\max}(\boldsymbol{A})$	Maximum eigenvalue of matrix $\boldsymbol{A}$
$\mathrm{diag}(\boldsymbol{A}_1, \ldots)$	Block-diagonal matrix with blocks $\boldsymbol{A}_1, \ldots$
$\lvert\lvert\boldsymbol{A}\rvert\rvert_2$	Induced 2-norm of matrix $\boldsymbol{A}$, i.e. $\lvert\lvert\boldsymbol{A}\rvert\rvert_2 = \sqrt{\lambda_{\max}(\boldsymbol{A}^T\boldsymbol{A})}$
$\lvert\lvert\boldsymbol{x}\rvert\rvert_2$	Euclidean norm of vector $\boldsymbol{x}$, i.e. $\lvert\lvert\boldsymbol{x}\rvert\rvert_2 = \sqrt{\boldsymbol{x}^T\boldsymbol{x}} = \sqrt{x_1^2 + \ldots + x_\mathrm{n}^2}$
$\lfloor x \rfloor$	Floor, i.e. $\lfloor x \rfloor$ is the largest integer smaller than or equal to $x \in \mathbb{R}$
$x \bmod y$	Modulo, i.e. $x \bmod y = x - y \left\lfloor \frac{x}{y} \right\rfloor$ with $x, y \in \mathbb{R}$
$\lvert\mathbb{M}\rvert$	Cardinality of the set $\mathbb{M}$

Others

$\mathbf{0}$	Zero matrix
$\boldsymbol{I}$	Identity matrix
$\begin{pmatrix} \boldsymbol{A} & * \\ \boldsymbol{B} & \boldsymbol{C} \end{pmatrix}$	Symmetric matrix $\begin{pmatrix} \boldsymbol{A} & \boldsymbol{B}^T \\ \boldsymbol{B} & \boldsymbol{C} \end{pmatrix}$
x^*	Variable x determined by optimization

Acronyms

ADC	Analog-to-Digital Conversion
CAN	Controller Area Network
CDMA	Code Division Multiple Access
CSMA	Carrier Sense Multiple Access
CSMA-CD	Carrier Sense Multiple Access with Collision Detection
CSMA-CR	Carrier Sense Multiple Access with Collision Resolution
DRE	Difference Riccati Equation
EBCS	Event-Based Control and Scheduling
ECU	Electronic Control Unit
FDMA	Frequency Division Multiple Access
GPRS	General Packet Radio Service
GSM	Global System for Mobile Communication
GUAS	Global Uniform Asymptotic Stability
IACS	Implementation-Aware Control and Scheduling
LMI	Linear Matrix Inequality
LIN	Local Interconnect Network
LQR	Linear-Quadratic Regulator
LTI	Linear Time-Invariant
MAC	Medium Access Control
MACA	Multiple Access with Collision Avoidance
NCS	Networked Control System
NP	Non-Deterministic Polynomial Time
OPP	Optimal Pointer Placement
PBCS	Predictive-Based Control and Scheduling
PCS	Periodic Control and Scheduling
RHCS	Receding-Horizon Control and Scheduling
TDMA	Time Division Multiple Access
TOD	Try-Once-Discard
UMTS	Universal Mobile Telecommunications System
WLAN	Wireless Local Area Network
ZOH	Zero-Order Hold

1 Introduction

1.1 Networked Control Systems

Revolutionary developments in computation and communication technologies have recently led to a novel paradigm for control, namely networked control systems (NCSs). NCSs, as shown in Figure 1.1, consist generally of *possibly* interacted multiple control loops that are closed over *possibly* multiple communication networks.

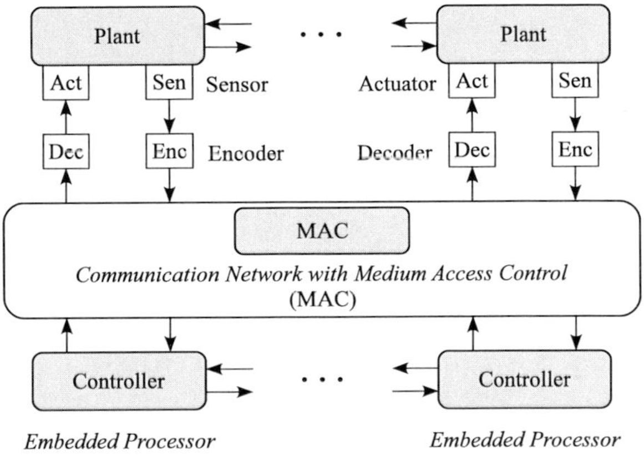

Figure 1.1: General architecture of a networked control system

NCSs provide several advantages from both cost and functionality perspectives over conventional hard-wired control systems. From the cost point of view, the advantages include

- *Installation and maintenance costs reduction.* Using a communication network instead of a hard-wired (point-to-point) connection between terminals (sensor, controller, actuator) reduces the wiring cost. Furthermore, fault detection and isolation of NCSs is relatively quick, easy, and more systematic.

- *Flexibility and reconfigurability improvement.* Closing a control-loop over a communication network increases the flexibility of adding to and removing from the network and also the reconfigurability of placing the terminals.
- *Efficient utilization of resources.* The communication network can be shared by a number of control loops larger than the number of available communication channels. Furthermore, a few number of embedded processors can be shared for executing control tasks of many control loops, utilizing them efficiently.

From the functionality point of view, the advantages of introducing a communication network within a control-loop include

- *Control of spatially distributed systems utilizing wide area networks.* Typical examples of spatially distributed systems include electric power networks [JHL+12], transportation networks [NSH08], irrigation networks [FMH+14], and manufacturing and process control [Lei09]. Conventional centralized control of such systems is usually neither feasible nor desirable since it leads to large and often very complex computational problems.
- *Control of mobile systems utilizing wireless networks.* Typical examples include control of autonomous mobile robots [D'A05] and autonomous ground vehicles [BLS07] as well as mobile sensor networks [OFL04]. In order to achieve a global objective, a communication between the mobile nodes (robots, vehicles, or sensors) is crucial.
- *Control based on multiple information.* All information that is sent over the utilized communication network such as sensor signals or control signals can be used for control. Thus, the resulting control performance due to this additional information can significantly be improved.

Due to these attractive advantages, considerable attention has been given in recent years to NCSs as witnessed by several surveys [Yan06, HNX07, Zam08, GC10, ZGK13], special issues [Bus01, AB04, AB07], and monographs [HVL05, BHJ10b, Lun14]. NCSs occur in various application domains, ranging from transportation systems like automobiles and aircrafts [JTN05], over manufacturing and process control [MT07], to power systems [LGOH09] and water distribution systems [LZNS10]. Typical motivating examples of some application domains are detailed in the following section.

1.2 Motivating Examples

The first example is taken from the transportation systems domain, particularly automobiles. Modern cars contain many of electronic control units (ECU) interconnected via fieldbuses (e.g. CAN, LIN, and FlexRay) and wireless (e.g. Bluetooth, ZigBee, and Wi-Fi) [NHB05]. Most of these units are used for control purposes such as vehicle dynamics control, engine control, cruise control, suspension control, climate control, and so on. We focus in the following on the suspension control unit with the architecture shown in Figure 1.2.

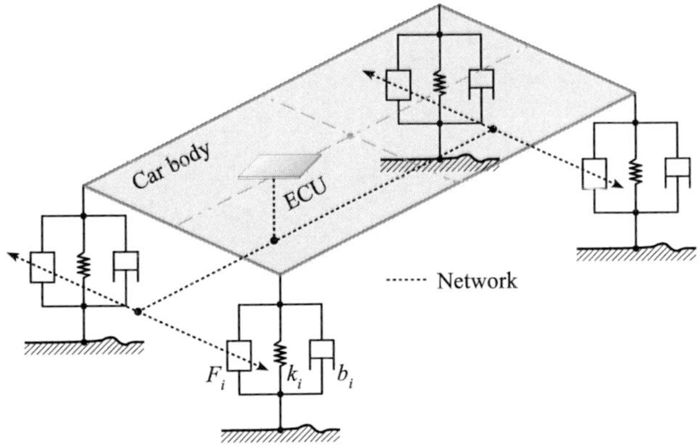

Figure 1.2: Architecture of an active suspension system

The active suspension system consists of four quarter cars connected by the car body (or sprung mass). Each quarter car is further composed of a spring, a shock absorber, and a hydraulic actuator. The control objective consists in improving ride comfort and road handling. The ride comfort is usually quantified by the sprung mass acceleration while the road handling performance by the force between car's tire and road surface. To achieve this control objective, a centralized ECU is used for coordinating the four actuators over a fieldbus (serial bus) network. The architecture of the active suspension system outlined above can be visualized based on the general architecture of NCSs shown in Figure 1.1 in a straightforward manner. The quarter cars can be seen as dynamically-coupled plants (or agents) controlled (or coordinated) over a wired communication network.

The second example (or testbed for developing new tools and techniques of controlling autonomous systems) is the RoboCup robot soccer team [D'A05]. RoboCup is the soccer world championship for autonomous robots. A robot team must actually perform a soccer game, incorporating various technologies such as design principles of autonomous agents, multi-agent collaboration, real-time reasoning, robotics, and sensor fusion. The physical architecture used by the Cornell RoboCup team (the winner of the competition four times in the time period 1999-2003) is depicted in Figure 1.3. The robot team as well as the other game objects are identified by a global vision system, with up to three cameras. The collected information is sent over a serial bus communication network (Ethernet) to a centralized control workstation for coordinating the motion of the

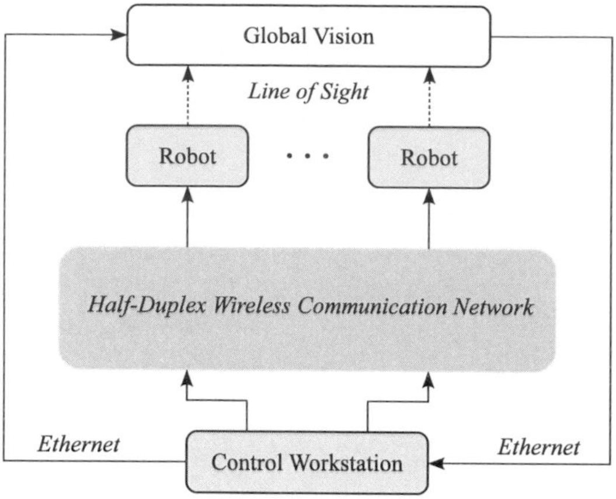

Figure 1.3: Physical architecture of the Cornell RoboCup team

robots. The control data are sent further to the mobile robots over a wireless communication network and to the cameras over a serial bus. The physical architecture depicted in Figure 1.3 can be described based on the NCS architecture shown in Figure 1.1. The mobile robots represent dynamically-decoupled plants (or agents) controlled (or coordinated) over a wireless communication network.

1.3 Network-Induced Imperfections

Although the advantages of NCSs are appealing, the utilization of a communication network within a control-loop induces different types of imperfections. These network-induced imperfections can be categorized in five types [HTWN10]:

Variable Transmission Times

The term *transmission time* refers to the time required for encapsulating data in packets at sender side, sending them over a physical medium, and releasing them at the destination side. According to the NCS architecture depicted in Figure 1.1, it does imply the transmission time τ^{SC} from sensor to controller and the transmission time τ^{CA} from controller to actuator. These transmission times might be uncertain and time-varying due to the MAC protocols, varying packet size, routing mechanism etc. Further details on the most commonly used MAC protocols are given in the following section.

From a control perspective, there are several ways of representing the transmission times proposed in literature. Firstly, representing them as uncertain time-varying but interval bounded time delays, see e.g. [HDL06, CWHN09, GOL+10, DHWH11], and ignoring any other additional information such as a possibly stochastic process characterizing their dynamics. Exceptions are the works proposed in [NBW98, DHB+12], in which a specific stationary probability distribution, namely an independent and identical distribution (i.i.d.), is assumed in addition. Secondly, representing them as stochastic time delays characterized by stochastic processes, e.g. a Markov process, but belonging to a finite set, see e.g. [ZSCH05, MC08, SY09]. Thirdly, representing them more generally as uncertain time delays that belong to a finite set of different bounded intervals (not a finite set of possible delays and not only one bounded interval as in the previous two methods), see [AGL14, AGL15a]. Furthermore, the transition from one bounded interval to another one can be either arbitrary or described by a stochastic process. Such representation is reasonable due to the fact that the network-induced transmission time is a function of the traffic load, number of nodes, and length of data packets which may vary considerably over time as shown in [PMFJ12].

Variable Sampling Instants

NCSs are generally spatially distributed systems in which a common notion of time reference of specified accuracy is usually provided via a clock synchronization protocol. Clock synchronization can be achieved in several ways such as software synchronization, hardware synchronization, or a combination of both. Several successful clock synchronization protocols have been developed for both wired and wireless communication networks, see e.g. the survey paper [SBK05]. Although the maximum deviation among the internal clocks within a set of synchronized nodes can reasonably be small, it might lead to uncertain time-varying sampling instants.

From a control perspective, there are different ways of representing the variable sampling instants proposed in literature. First, representing them implicitly as uncertain time-varying time delays that are reset at actuation or sampling instants, see e.g. [FSR04, Mir07, NHT08]. Second, representing them explicitly as uncertain time-varying sampling intervals that belong to a finite set as in [MA03] or to a bounded interval as in [Suh08, Fuj09, CWHN09].

Medium Access Constraints

Access constraints result from sharing the communication network medium by a number of nodes larger than the number of available communication channels. From a control perspective, two different directions have been proposed in literature for representing this imperfection. First, representing it implicitly as a time delay that might be uncertain and time-varying depending on the utilized MAC protocol, see e.g. [NBW98, DN07, NH09, HTWN10]. Second, representing it explicitly as an additional constraint that is incorporated with system dynamics, see e.g. [RS04, GHQW04, CA06, GIL09] for a single-channel and [ZHV06, BÇH06, DLG10] for a multi-channel case.

Packet Dropouts

Communication networks can usually be viewed as unreliable data transmission paths, where packet collision, network node failure, and congestion occasionally occur. In such cases, it might be advantageous from a control perspective to drop the old packet and transmit a new one instead of repeated retransmission attempts [BA12a]. Moreover, disordering of packets due to routing may lead to outdated and thus lost packets as well.

Packet dropouts imperfection has been represented in literature differently. It has been implicitly represented as a prolongation of time delays, see e.g. [CWHN09, GLS$^+$12], a prolongation of sampling intervals, see e.g. [GRB07], or explicitly as in [SDHW10] with an upper bound on the number of consecutive dropped packets. In case that the maximum number of consecutive dropped packets is not known, packet dropouts have been represented as a stochastic process such as a Bernoulli process [SSF$^+$07] or a Markov process [WC07, GHBW12]. Representing the packet dropouts phenomenon as a stochastic process while taking the maximum number of consecutive dropped packets into account has been represented in [BB09].

Quantization Errors

Quantization errors refer to the difference between an actual analog value and its quantized digital value in case of analog-to-digital conversion (ADC). Such errors occur as a result of the limited resolution of the utilized converters. Within the sampled-data control framework, much effort has been expended in studying the effect of quantization errors on the behavior of the control system, see e.g. [FPW90, Chapter 7].

Communication networks with data rate constraints necessitate the utilization of ADCs with low resolution, making the effect of quantization much more pronounced [HOV02]. From a control perspective, quantization errors are represented as an additional noise or perturbation $\boldsymbol{d}(t)$ added to the system [Lib03b, Section 5.3]. Based on the type of the utilized quantizer, different upper bounds on the Euclidean norm of the additional noise $\|\boldsymbol{d}(t)\|_2$ can be obtained. In case of static state quantization using uniform quantizers [HSJW03, LDWH12], a constant upper bound proportional to the maximal size of finite quantization regions is obtained. Under logarithmic quantizers [GC08, YSLG11], a time-varying upper bound that belongs to a finite set is rather obtained. Under dynamic state quantization [BL00, HOV02, Lib03a], finally, a time-varying upper bound that belongs to an infinite set is obtained.

1.4 MAC Protocols for NCSs

One of the most important network-induced imperfections outlined above is the medium access constraints. This imperfection has been tackled in practice by using a medium access control (MAC) protocol [TW11, Chapter 4]. MAC protocols, known also as schedulers, are aimed at distributing the limited number of available communication channels among the competing nodes according to some criterion. MAC protocols can be classified based on the utilized criteria as:

Contention-Free MAC Protocols

In contention-free MAC protocols, the connected nodes are pre-coordinated to use shared communication channels such that any contention among them is completely avoided. The pre-coordination can be based on time, frequency or space, leading to the following widely used MAC protocols:

- Time Division Multiple Access (TDMA)
- Frequency Division Multiple Access (FDMA)
- Code Division Multiple Access (CDMA)
- Token Passing or Master-Slave (Polling)

The basic idea of the TDMA protocol is rather simple: Each node is assigned a specific time slot in a round-robin fashion for getting the entire bandwidth. This is graphically illustrated in Figure 1.4. Such a protocol has been applied in several applications including industrial wireless networks such as WirelessHART and ISA-100 [JJ10]. Besides its simplicity, it allows a dynamic assignment of the time slots on demand. Moreover, each node needs to listen and broadcast only for its own time slot. For the rest of the time, other tasks can be carried out or be idle. Although the advantages of the TDMA protocol are interesting, guard times between time slots might limit the potential bandwidth. Furthermore, hard limits must apply on the network size to ensure that transmitted data is received at precisely the right time.

The FDMA protocol is so named because the entire bandwidth is divided into several frequency bands each assigned permanently to a node. It has been applied to numerous applications, including broadband wireless networks (WiMax) and power line networking [AGM07]. It is algorithmically simple since it does not require synchronization or timing control compared to the TDMA protocol. Moreover, it is not sensitive to near-far problem which is pronounced for the CDMA protocols. On the other hand, guard bands

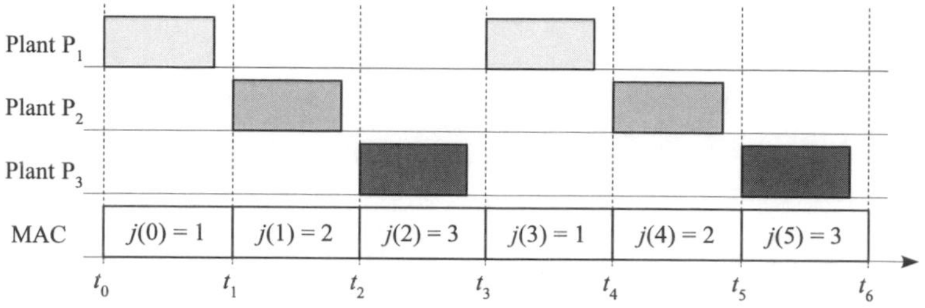

Figure 1.4: An illustration of the TDMA protocol. The output of the scheduler (MAC) is denoted by $j(k)$, $\forall k \in \mathbb{N}_0$

are required between the frequency bands to minimize adjacent channel interference, leading to unused frequency slots. Moreover, it necessitates high-performing filters in the radio hardware, in contrast to TDMA and CDMA protocols.

Under CDMA protocols, each node is assigned a specific orthogonal/pseudorandom code for getting the entire bandwidth all the time. It occurs in various applications including satellite navigation systems and wireless control with Bluetooth [VD05]. Although there is no strict limit on the number of nodes that can be supported, multiple access interference due to a partial rejection of unwanted signals may limit its capacity. Furthermore, power levels of transmitting nodes must be controlled to minimize the resulting multiple access interference.

In Token Passing protocols, a free data frame (token) is circulating on the network in a predefined order. If a node does not have data to be sent, it simply passes the token to the next station. Otherwise, it captures the free token and replaces it with its own message. A node may continue transmitting data until a timer (token holding timer) expires. This protocol has been used in computer network standards such as IEEE 802.5 and IEEE 802.17 and in industrial wired networks such as Profibus and ControlNet [LMT01]. The good thing in this protocol is its ability to handle a very heavy load better than well-known protocols like Ethernet for instance. It costs, however, several times as much as Ethernet hardware. Master-Slave polling can be viewed as a centralized version of the Token Passing protocol.

Contention-Based MAC Protocols

Due to the static channel allocation scheme of the contention-free MAC protocols outlined above, network-induced latency is (under perfect channel conditions) deterministic. They tend to work best when the traffic load (number of nodes and their traffic pattern) is predictable. On the other hand, not all nodes want to send (i.e. active) all the time and the number of active nodes is usually unpredictable. Furthermore, the active nodes should get full channel bandwidth for higher throughput. All of these arguments have motivated the prominence of contention-based MAC protocols in which communication channels are allocated dynamically without any pre-coordination. Since there is no pre-coordination, the nodes might compete for accessing shared channels, leading to a collision. According to how the collision is resolved, contention-based MAC protocols can be classified as

- Collision recovery (e.g. ALOHA, Slotted ALOHA, and Reservation ALOHA)
- Collision avoidance (e.g. CSMA, CSMA-CD, and MACA)
- Collision free (e.g. CSMA-CR)

Collision does matter since it degrades channel utilization and leads to uncertain time-varying transmission times and packet dropouts. Such factors might prohibit its application in real-time traffic.

The main idea of collision recovery MAC protocols is to let nodes transmit whenever they have data to be sent. If a collision occurs (two or more nodes transmit at the same time), the collided data will be damaged. The concerned nodes have to wait a random amount of time (based on a backoff algorithm) and then resend their data again. The waiting time must be random, otherwise the same nodes will collide over and over. These protocols have been adapted for use in all major mobile telephone standards such as GSM, GPRS, and UMTS [BA02, Chapter 3]. The concept is very simple and applicable to any network in which uncoordinated nodes are competing for the use of shared communication channels. It leads however to low efficiency (poor channel utilization) in general.

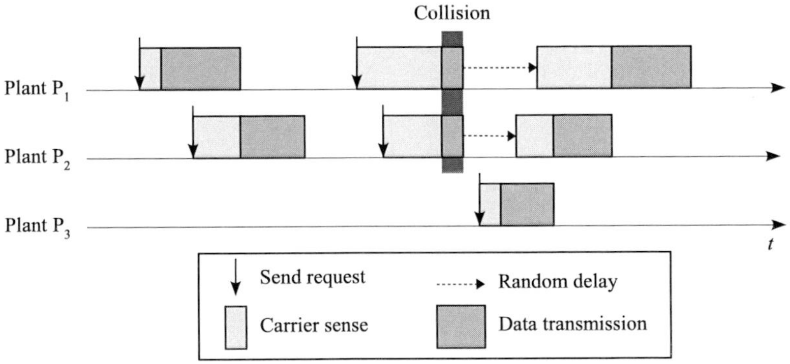

Figure 1.5: A graphical illustration of the CSMA protocol

Collision avoidance MAC protocols can be viewed as improved versions to the collision recovery ones. When a node has data to send, it first listens to the channel. If the channel is idle, the node can send its data. Otherwise it just waits until the channel becomes idle as shown in Figure 1.5 for the CSMA protocol. Listening-before-sending reduces the chance of collision, but it does not fully solve the collision problem. If a collision occurs because of two or more listening nodes simultaneously deciding to transmit, the same procedure as for the collision recovery protocols is followed. The applications of these MAC protocols include Hub-based Ethernet, indoor WLAN, and industrial wireless networks such as ZigBee [JJ10].

Collision free protocols are extended versions to the collision avoidance ones. If a node wishes to send a message, it sends its identifier bit by bit starting from the most significant one. If a collision happens, a bitwise arbitration mechanism is used to resolve it. More precisely, bits from different nodes are BOOLEAN ORed/ANDed together by the channel. The winner (higher priority) node can now send its message without collision while losers switch to the receiving mode. This concept has been applied in CAN-based networks such as DeviceNet, CANopen, CAN kingdom, and TTCAN [JTN05]. It is simple, elegant, efficient, and deterministic in nature. However, it does not find a home yet in wireless networks. Moreover, the bit arbitration mechanism is sensitive to medium properties. Thus, the maximum length of a wired network is limited.

1.5 Modeling and Analysis of NCSs

For stability analysis and control synthesis of NCSs, a mathematical model that captures besides system dynamics the network-induced imperfections outlined above is of particular importance. Different directions of modeling and analysis of NCSs have been proposed in the literature, see [BHJ10a] for an overview. They are categorized in the following under a couple of headers. Although the categories are not exact, they should give a flavor of what methods available today and what the main differences are. For the sake of exposition, we focus in the following on linear systems to describe the main idea and the properties of each of them.

Continuous-Time Approach

In this approach, the network-induced imperfections including variable transmission times, variable sampling instants, medium access constraints, and packet dropouts have been implicitly represented as uncertain time-varying input delays [FSR04, YWCH05, HBSW07, GCL08]. More precisely, consider a linear sampled-data (due to the digital implementation platform) system

$$\dot{\boldsymbol{x}}(t) = \boldsymbol{A}\boldsymbol{x}(t) + \boldsymbol{B}\boldsymbol{u}(t_k) \qquad \forall t \in [t_k + \tau_k, t_{k+1} + \tau_{k+1}[\tag{1.1}$$

where $\boldsymbol{x}(t) \in \mathbb{R}^n$ is the state vector, $\boldsymbol{u}(t) \in \mathbb{R}^m$ is the control vector, $\tau_k \in \mathbb{R}_0^+$ is the variable transmission time, and $t_k \in \mathbb{R}_0^+$ is the k-th sampling instant with the subindex

$k \in \mathbb{N}_0$. Equation (1.1) can equivalently be rewritten as

$$\begin{aligned} \dot{\boldsymbol{x}}(t) &= \boldsymbol{A}\boldsymbol{x}(t) + \boldsymbol{B}\boldsymbol{u}\big(t - \tau(t)\big) \\ \tau(t) &= t - t_k \end{aligned} \qquad \forall t \in [t_k + \tau_k, t_{k+1} + \tau_{k+1}[\tag{1.2}$$

with the uncertain time-varying input delay $\tau(t) \in [\tau_{\min}, \tau_{\max}]$ representing the network-induced imperfections outlined above.

Based on the NCS model (1.2), closed-loop stability as proposed in [YHP04] has been studied using the Lyapunov-Krasovskii functionals (a generalization of the Lyapunov direct method [Ric03]) such as

$$V(t) = \boldsymbol{x}^T(t)\boldsymbol{P}\boldsymbol{x}(t) + \int_{t-\tau_{\max}}^{t} \int_{s}^{t} \dot{\boldsymbol{x}}^T(v)\boldsymbol{T}\dot{\boldsymbol{x}}(v)dvds \tag{1.3}$$

with the symmetric and positive definite matrices $\boldsymbol{P}, \boldsymbol{T} \in \mathbb{R}^{n \times n}$. Alternatively, the Lyapunov-Razumikhin function, a particular case of the Lyapunov-Krasovskii one which generally leads to conservative results but can directly be used for control synthesis, can be used [Ric03]. A typical example of such functions, see e.g. [YWCH04], is given by

$$V(t) = \boldsymbol{x}^T(t)\boldsymbol{P}\boldsymbol{x}(t) \tag{1.4}$$

with the symmetric and positive definite matrix $\boldsymbol{P} \in \mathbb{R}^{n \times n}$.

Remark 1.1. Within the continuous-time framework, quantization errors can also be considered in the form of an additive and bounded disturbance, see [FD09] for further details.

Discrete-Time Approach

In this approach, an exact discrete-time representation of the linear sampled data system (1.1) over the sampling interval $t_k \leq t < t_{k+1}$ using ZOH is first constructed [CWHN06, HDL07, IGL10, DHWH11]. Due to the network-induced imperfections outlined above, we end up with a discrete-time switched linear system

$$\boldsymbol{x}(k+1) = \boldsymbol{A}_{j(k)}(\tau_k, h_k)\boldsymbol{x}(k) + \boldsymbol{B}_{j(k)}(\tau_k, h_k)\boldsymbol{u}(k) \tag{1.5}$$

with the indices $k \in \mathbb{N}_0$ denoting, for notational convenience, the sampling instants $t_k \in \mathbb{R}_0^+$. For a general discussion on switched systems, the reader can refer e.g. to survey papers [SG05a, LA09] or books [Lib03b, SG05b]. The switching index $j(k) \in \mathbb{M}$, where the finite set $\mathbb{M}$ follows from the architecture of the utilized communication network, is employed to explicitly represent scheduling due to medium access constraints. For uncertain but interval bounded time-varying transmission times τ_k and sampling intervals $h_k \triangleq t_{k+1} - t_k$, the parameter variation of the resulting switched system model

(1.5) belongs to an infinite set. To enable the application of the well-known robust control theory [KBM96, DB01], overapproximation or embedding techniques, cf. Section 2.3, are next applied to (1.5), leading in general to a discrete-time switched polytopic system with additive norm-bounded uncertainty

$$\boldsymbol{x}(k+1) = \left(\sum_{\ell=0}^{L} \mu_\ell(k)\boldsymbol{A}_{j(k)\ell} + \Delta\boldsymbol{A}_{j(k)}\right)\boldsymbol{x}(k) + \left(\sum_{\ell=0}^{L} \mu_\ell(k)\boldsymbol{B}_{j(k)\ell} + \Delta\boldsymbol{B}_{j(k)}\right)\boldsymbol{u}(k) \quad (1.6)$$

with the uncertain parameters $\mu_\ell(k) \in \mathbb{R}_0^+$ such that $\sum_{\ell=0}^{L} \mu_\ell(k) = 1$, the polytopic vertices $\boldsymbol{A}_{j(k)\ell}$ and $\boldsymbol{B}_{j(k)\ell}$, and the norm-bounded uncertainties $\Delta\boldsymbol{A}_{j(k)}$ and $\Delta\boldsymbol{B}_{j(k)}$.

Based on the resulting NCS model (1.6), stability analysis as well as control synthesis under arbitrary/constrained switching can be studied using multiple Lyapunov functions [Bra98, DBPL00, DB01, DRL02]. An example of such functions is a switched parameter-dependent Lyapunov function [HDL06, AGL11]

$$V(k) = \boldsymbol{x}^T(k)\left(\sum_{\ell=0}^{L} \mu_\ell(k)\boldsymbol{P}_{j(k)\ell}\right)\boldsymbol{x}(k) \quad (1.7)$$

with the symmetric and positive definite matrices $\boldsymbol{P}_{j(k)\ell} \in \mathbb{R}^{n\times n}$ for all $j(k) \in \mathbb{M}$ and $\ell \in \{0, \ldots, L\}$. Typical examples of constrained switching include minimum/average dwell time [GC06b, SQ10] or periodicity [RS04, ZHV06]. These constraints can be considered during the stability analysis and/or control synthesis, reducing thus the induced conservatism significantly.

Remark 1.2. By representing the packet dropouts imperfection implicitly as a prolongation of transmission times or sampling intervals, it can easily be included in the NCS model (1.5). Furthermore, certain extensions can be made to the above setup such that quantization errors are accounted for as shown in [YSLG11, LDWH13].

Remark 1.3. Although an exact discretization can be obtained only for linear sampled-data systems, the discrete-time approach outlined above can also be applied to NCSs with nonlinear dynamics as discussed in [WNH12]. Moreover, at least for the case of NCSs with linear dynamics, it has been shown in [DHWH11] that the discrete-time approach leads to less conservative stability results compared to the hybrid approach which is discussed in the following.

Remark 1.4. Alternative to the discrete-time switched linear system (1.5), the NCS can also be modeled as a continuous-time switched linear system [LZFA05, GC06a, CGA08]

$$\dot{\boldsymbol{x}}(t) = \boldsymbol{A}_{j(k)}\boldsymbol{x}(t) + \boldsymbol{B}_{j(k)}\boldsymbol{u}(t). \quad (1.8)$$

Closed-loop stability of the NCS is further analyzed based on the average dwell time approach together with multiple Lyapunov functions.

Hybrid Approach

The term *hybrid* is used here to refer to systems involving interactions between continuous dynamics (flow) and discrete dynamics (jump) as used for the first time in [Wit66]. The flow part is described by differential/difference equations while the jump part by logics. In this approach, neither a discretization of the linear sampled-data system (1.1) nor reformulation as a continuous-time system has been exploited. The linear sampled-data system (1.1) is left unchanged, i.e.

$$\begin{aligned} \dot{\boldsymbol{x}}(t) &= \boldsymbol{A}\boldsymbol{x}(t) + \boldsymbol{B}\boldsymbol{u}(t_k) \\ \boldsymbol{u}(t_k) &= \boldsymbol{K}\boldsymbol{x}(t_k) \qquad \forall t \in [t_k + \tau_k, t_{k+1} + \tau_{k+1}[\end{aligned} \tag{1.9}$$

with the static control gain $\boldsymbol{K} \in \mathbb{R}^{m\times n}$. Introducing an additional state variable

$$\boldsymbol{z}(t) \triangleq \boldsymbol{x}(t_k),$$

the system (1.9) can equivalently be rewritten as a hybrid (impulsive) system [NHT07]

$$\begin{pmatrix} \dot{\boldsymbol{x}}(t) \\ \dot{\boldsymbol{z}}(t) \end{pmatrix} = \begin{pmatrix} \boldsymbol{A} & \boldsymbol{BK} \\ \boldsymbol{0} & \boldsymbol{0} \end{pmatrix} \begin{pmatrix} \boldsymbol{x}(t) \\ \boldsymbol{z}(t) \end{pmatrix} \qquad \forall t \in [s_k, s_{k+1}[\tag{1.10a}$$

$$\begin{pmatrix} \boldsymbol{x}(s_{k+1}) \\ \boldsymbol{z}(s_{k+1}) \end{pmatrix} = \begin{pmatrix} \boldsymbol{x}(s_{k+1}) \\ \boldsymbol{x}(t_{k+1}) \end{pmatrix} \qquad \forall k \in \mathbb{N}_0 \tag{1.10b}$$

where $s_k = t_k + \tau_k$ is the k-th control update instant. The network-induced imperfections such as the variable transmission time τ_k and sampling interval h_k are assumed to be uncertain but interval bounded, i.e.

$$\tau_k \in [\tau_{\min}, \tau_{\max}] \quad \text{and} \quad h_k \in [h_{\min}, h_{\max}]. \tag{1.11}$$

Based on the resulting NCS model (1.10), closed-loop stability has been studied using extended versions of the Lyapunov-Krasovskii functional (1.3), see [NHT07, NHT08] for further details.

Remark 1.5. Packet dropouts imperfection can be accounted for in the hybrid approach by representing it implicitly as a prolongation of transmission times or sampling intervals. Moreover, medium access constraints and quantization errors can be included as discussed in [NH09, NL09, HNTW09, HTWN10, LFH12]

Remark 1.6. The hybrid approach has been mainly proposed for NCSs with nonlinear dynamics with which the linear sampled-data NCS outlined above can be considered as a special case [NT04a, NT04b]. This allows its applicability to a wide class of networked control systems. Moreover, it has been shown in [NHT08] that the proposed stability conditions within the hybrid framework are always either less or equally conservative compared to the proposed ones within the continuous-time framework mentioned before.

1.6 Control and Scheduling Codesign

Several MAC protocols have been developed, as discussed before, including contention-free protocols such as TDMA or FDMA and contention-based protocols such as CSMA or any of its versions. The main problem with such MAC protocols is that control requirements are generally not considered and consequently the resulting control performance is usually low due to either the static nature of the contention-free protocols or the stochastic nature of the contention-based protocols. Therefore, much research in recent years has focused on control and scheduling codesign for NCSs, see e.g. the recent survey paper [YWL+11, WS12], rather than a separation of concerns in which control and scheduling are addressed separately under some *ideal* assumptions like constant sampling intervals or *conservative* assumptions like worst-case transmission times.

It is worth to note that the control and scheduling codesign concept has also been used in embedded control systems where a set of control tasks are implemented on a microprocessor embedded into the application with a shared CPU-time [ÅCES00, BPZ02, ÅC05, XS06, LVM07, XS08, SCEP09]. Well-known *preemptive* scheduling policies such as the rate-monotonic (RM) for fixed priorities and the earliest-deadline first (EDF) for dynamic priorities have been designed based on timing parameters such as sampling interval, worst-case execution time, and deadline. Meeting deadlines does not however imply constant sampling periods and latencies as illustrated in [GIL07]. Resulting sampling periods and latencies under such policies are subject to variations. Moreover, control requirements based on performance specifications do not fit well to schedulability tests in which the objective is just to maximize the number of tasks to be scheduled. Consequently, a blind use of such classical policies might lead to a poor control performance or even instability of the system under consideration [CHL+03].

Control and scheduling codesign is thus very promising for achieving high control performance under limited resources. This is generally attributed to using dynamic (not static) and deterministic (not stochastic) scheduling algorithms, on the one hand, and explicitly regarding them during the control design, on the other hand. Dynamism is required for distributing shared resource according to actual demands and not based on a predefined scenario while determinism for rigorous certificate of system stability. Such scheduling properties besides the implementation-aware control design are the stimuli behind addressing control and scheduling codesign. Various concepts for control and scheduling codesign have been proposed in literature. They can roughly be classified into two categories, namely *time-triggered* codesign and *event-triggered* codesign.

Time-Triggered Codesign

Under time-triggered codesign, scheduling decision and control update instants are coordinated by a clock. Time-triggered codesign approaches can further be classified from the scheduling perspective into static and dynamic scheduling. Under static scheduling, the schedule is determined before runtime (offline), where usually periodic schedules are considered [RS00, RS04, ZHV06, XJ08, LHB09, BÇH09]. Besides the medium access constraints, the variable transmission time imperfection has been addressed in [ZLR08, DLG10]. The variable transmission time has been represented as an uncertain but interval bounded time-varying input delay. The resulting codesign problem has been formulated as an LMI optimization problem using a common quadratic Lyapunov function in [ZLR08] or using the average dwell-time technique in [DLG10]. Generally, the main property of the aforecited time-triggered codesign strategies with static scheduling is that the required scheduling overhead is low at the cost of no adaptation under disturbances. To allow online reaction to disturbances, dynamic scheduling is indispensable.

Under dynamic scheduling, the schedule is determined at runtime (online) based on measurements of system's state at each sampling instant [CA06, BÇH06, GIL09, GIL11]. The proposed codesign strategies in [CA06, BÇH06, GIL09, GIL11] require state measurements from all plants to be sent to the centralized scheduler at each sampling instant for decision making. This requirement may lead to a large amount of scheduling overhead in NCSs which in turn may counteract the reactivity property of dynamic schedules to disturbances. Alternatively, the decentralized try-once-discard (TOD) scheduling policy, cf. Remark 7.3, proposed in [WYB02] can be used, as discussed in [DN07]. The aforecited codesign strategies with dynamic scheduling are characterized by considering only the access constraints imperfection while neglecting the other network-induced imperfections mentioned above.

Event-Triggered Codesign

Under event-triggered codesign, scheduling decision and control update instants are coordinated by an event generator rather than a clock. Event-triggered control/scheduling has been suggested in literature as a way to improve the control performance compared to the time-triggered counterpart using the same update rate. Various papers on event-triggered control/scheduling are listed and classified based on types of network-induced imperfections under consideration in Table 1.1. From Table 1.1, one can notice that event-triggered control/scheduling has been extensively studied over the past ten years. Event-triggered control and scheduling codesign has however been rarely addressed in literature, see [CH08, RJ09, MH11, PTNA11, RSJ13]. Few concepts have been developed yet for specific scenarios such as a class of first-order plants [CH08, RJ09], using the separation principle (certainty equivalence) [MH11, RSJ13], or using a centralized event-triggering policy [PTNA11]. Noteworthy, the term *event-triggered codesign* is occasionally used in literature to refer to the joint design of an event-triggering policy and a controller for single plants, see e.g. [MH09, SB11, PY13, MC14]. The triggering policy is called with abuse of words a *scheduler*, creating some confusion with the framework considered in this thesis in which multiple plants (and not a single plant) are considered.

Paper	Trans. time	Sampl. instant	Access constr.	Packet dropout	Quanti. errors
[Årz99, Åst08, HJT12, LWHZ14]	–	–	–	–	–
[LL11a, LH13, Dur13]	✓	–	–	–	–
[CJ08]	✓	✓	–	–	–
[Tab07, CH08*]	✓	–	✓	–	–
[WL11, GLS+12, DFH13, FHDH13]	✓	–	–	✓	–
[YA13, GM09]	✓	–	–	–	✓
[RSBJ11, BA11, HQSJ12, RSJ13*]	–	–	✓	–	–
[RJ09*, MH11*]	–	–	✓	✓	–
[MH10, BA12b]	–	–	–	✓	–

Table 1.1: Classification of papers on event-triggered control and/or scheduling. Papers that address event-triggered codesign are highlighted by superscript asterisks

1.7 Objectives and Contributions

According to the short overview outlined above, one can easily notice that control and scheduling codesign for NCSs is in its infancy. For event-triggered codesign, especially, few concepts and methods have been proposed in recent years. Control and scheduling codesign for NCSs while taking the network-induced imperfections explicitly into account has been rarely addressed. This thesis aims at filling this gap by unifying and generalizing existing methods as well as providing novel methods. To this end, new concepts and techniques for analysis and design are in order. The objectives and contributions of the thesis are detailed in the following.

Within this thesis, we consider a set of *independent* plants P_i, $\forall i \in \{1, \ldots, M\}$, each described by a continuous-time LTI state equation. Following the discrete-time approach in Section 1.5, the resulting NCS model including the network-induced imperfections is given by (1.6). As a structure of the controller, we consider further a state feedback control law

$$\boldsymbol{u}(k) = \boldsymbol{K}(k)\boldsymbol{x}(k) \tag{1.12}$$

where $\boldsymbol{K}(k)$ is a time-varying feedback matrix to be determined. For evaluation purposes, we consider finally a quadratic cost function

$$J = \sum_{k=0}^{\infty} \begin{pmatrix} \boldsymbol{x}(k) \\ \boldsymbol{u}(k) \end{pmatrix}^T \boldsymbol{Q}_{j(k)} \begin{pmatrix} \boldsymbol{x}(k) \\ \boldsymbol{u}(k) \end{pmatrix} \tag{1.13}$$

where $\boldsymbol{Q}_{j(k)}$ is a switched positive semidefinite weighting matrix. The control objective is summarized as follows

Problem 1.1 *For the NCS model* (1.6) *and the control law* (1.12) *find the optimal feedback sequence* $\boldsymbol{K}^*(0), \ldots, \boldsymbol{K}^*(\infty)$ *and switching sequence* $j^*(0), \ldots, j^*(\infty)$ *such that the cost function* (1.13) *is robustly minimized, i.e.*

$$\min_{\substack{j(0),\ldots,j(\infty)\\ \boldsymbol{K}(0),\ldots,\boldsymbol{K}(\infty)}} \max J \qquad \text{subject to (1.6) and (1.12).} \tag{1.14}$$

Five different methods to tackle Problem 1.1 are studied in this thesis. The main idea behind each of them along with a short review on the related work is given in the following.

Periodic Control and Scheduling

In this strategy, it is assumed that the switching sequence is p-periodic, i.e.

$$j(k+p) = j(k) \qquad \forall k \in \mathbb{N}_0. \tag{1.15}$$

Based on this assumption, the corresponding p-periodic feedback sequence

$$\big(\boldsymbol{K}(k), \ldots, \boldsymbol{K}(k+p-1)\big) \tag{1.16}$$

is determined from an LMI optimization problem using a p-periodic Lyapunov function [DT00]. This is repeated for all admissible p-periodic switching sequences, leading to the set $\mathbb{S}_p$ of p-periodic switching-feedback sequences. So far, everything is done completely offline. Online, the optimal switching index $j^*(k)$ together with the optimal feedback matrix $\boldsymbol{K}^*(k)$ are determined at time instant t_k for the current state $\boldsymbol{x}(k)$ based on exhaustive search, relaxation, or optimal pointer placement. Under exhaustive search, all elements of the set $\mathbb{S}_p$ are inspected at each time instant for determining the optimal switching index $j^*(k)$ and the corresponding feedback matrix $\boldsymbol{K}^*(k)$. Although closed-loop stability is guaranteed inherently under exhaustive search, the resulting online complexity is characterized by $|\mathbb{S}_p| \cong M^p$ that grows exponentially with the period p. In order to reduce the online complexity, a relaxed set $\hat{\mathbb{S}}_p \subset \mathbb{S}_p$ is extracted provided that closed-loop stability is preserved, as discussed in [Gör12]. Alternatively, the optimal pointer placement strategy proposed in [BÇH06, BÇH09] in which the relaxed set $\hat{\mathbb{S}}_p$ containing only one p-periodic element (determined offline by optimization) can be considered. The periodic control and scheduling codesign strategy proposed in this thesis can be viewed as a generalization of the work proposed in [BÇH06, BÇH09, Gör12] for switched linear systems (1.5) toward switched polytopic systems with additive norm-bounded uncertainty (1.6). The effectiveness of the proposed strategy and the effect of the chosen period p on the resulting control performance are evaluated by simulation for an illustrative example and also experimentally for a case study.

Receding-Horizon Control and Scheduling

Instead of imposing periodicity on the infinite switching sequence $j(0),\dots,j(\infty)$, we decompose the cost function (1.13) into two parts, i.e.

$$J = \underbrace{\sum_{k=0}^{N-1} \begin{pmatrix} \boldsymbol{x}(k) \\ \boldsymbol{u}(k) \end{pmatrix}^T \boldsymbol{Q}_{j(k)} \begin{pmatrix} \boldsymbol{x}(k) \\ \boldsymbol{u}(k) \end{pmatrix}}_{J_1} + \underbrace{\sum_{k=N}^{\infty} \begin{pmatrix} \boldsymbol{x}(k) \\ \boldsymbol{u}(k) \end{pmatrix}^T \boldsymbol{Q}_{j(k)} \begin{pmatrix} \boldsymbol{x}(k) \\ \boldsymbol{u}(k) \end{pmatrix}}_{J_2} \tag{1.17}$$

where $N \in \mathbb{R}^+$ is the prediction horizon. In the second part J_2, a p-periodic switching sequence $j(N), j(N+1), \dots, j(N+p-1)$ with p denoting the period is considered. Thus, the cost function (1.17) can be reformulated as a finite time-horizon quadratic cost function

$$J_N = \boldsymbol{x}^T(N)\boldsymbol{Q}_0\boldsymbol{x}(N) + \sum_{k=0}^{N-1} \begin{pmatrix} \boldsymbol{x}(k) \\ \boldsymbol{u}(k) \end{pmatrix}^T \boldsymbol{Q}_{j(k)} \begin{pmatrix} \boldsymbol{x}(k) \\ \boldsymbol{u}(k) \end{pmatrix} \tag{1.18}$$

where the terminal cost $\boldsymbol{x}^T(N)\boldsymbol{Q}_0\boldsymbol{x}(N)$ represents an upper bound on the cost from $k = N$ to $k = \infty$ using the p-periodic switching sequence. Based on the cost function (1.18), the control and scheduling codesign problem can be solved using dynamic programming [Bel57, Ber05]. The main idea is to start from the last prediction step $k = N-1$ and going backwards until the first prediction step $k = 0$. At each prediction step $k \in \{0,\dots,N-1\}$, the optimal feedback matrix $\boldsymbol{K}^*(k)$ is determined by an LMI optimization problem with the objective of minimizing an upper bound on the cost-to-go from k to N while the optimal switching index $j^*(k)$ by explicit enumeration. The resulting complexity under dynamic programming is characterized by the number of possible switching sequences M^N that grows exponentially with the prediction horizon N. The relaxed dynamic programming introduced in [LB02,LR06,Ran06] can instead be used to reduce the computational complexity while introducing further suboptimality. In order to suppress the effect of the imposed periodicity on the second part of the cost (1.17) and hence to improve the resulting performance, a reoptimization for the current state $\boldsymbol{x}(k)$ over all possible switching sequences is performed at each time instant t_k. This procedure is called receding-horizon control and scheduling and was introduced in [CA06] yet for an infinite prediction horizon, i.e. $N = \infty$, and extended in [GIL11] for finite prediction horizon. Both of them have however considered switched linear systems (1.5) with which a difference Riccati equation (DRE) [Ber05] for determining the optimal feedback matrix $\boldsymbol{K}^*(k)$ can be applied. Moreover, closed-loop stability has not been studied in [CA06] while a posteriori stability conditions have been proposed in [GIL11]. Extending their work toward switched polytopic systems with additive norm-bounded uncertainty (1.6) requires new analysis and design tools. Instead of using DREs, an iterative LMI optimization problem is proposed in this thesis. A novel stability condition is further proposed which can be checked a priori, i.e. before solving the iterative LMI optimization problem. The effectiveness of the proposed strategy and the effect of the prediction horizon N on the resulting performance are evaluated by simulation for an illustrative example and also experimentally for a case study.

Implementation-Aware Control and Scheduling

In this strategy, it is assumed during the codesign process that the optimal switching index $j^*(k)$ is determined for the current state $\boldsymbol{x}(k)$ based on a state-based switching law

$$j^*(k) = \arg \min_{j(k)\in\mathbb{M}} \boldsymbol{x}^T(k)\boldsymbol{P}_{j(k)}\boldsymbol{x}(k) \tag{1.19}$$

where $\boldsymbol{P}_{j(k)}$ is a switched symmetric and positive definite matrix. The state-based switching law (1.19) is introduced in [GC06b,GC06a] for stabilizing autonomous switched systems. Based on the state-based switching law (1.19), Problem 1.1 is transformed into a codesign problem of the switched matrix $\boldsymbol{P}_{j(k)}$ and the corresponding switched feedback matrix $\boldsymbol{K}_{j(k)}$ for all $j(k) \in \mathbb{M}$. The resulting codesign problem is formulated as an LMI optimization problem using Lyapunov-Metzler functions to be solved online. The optimal feedback matrix at time instant t_k is thus given by

$$\boldsymbol{K}^*(k) = \boldsymbol{K}_{j^*(k)} \tag{1.20}$$

with the optimal switching index $j^*(k)$ determined according to (1.19). Feasibility of the online LMI optimization problem as well as stability of the closed-loop system are analyzed. A more conservative version of the proposed codesign strategy that can be solved completely offline is further provided. The effectiveness of both the online and the offline version are evaluated by simulation for an illustrative example and also experimentally for a case study.

Event-Based Control and Scheduling

The three codesign strategies proposed above require transmission of state measurements from all plants at each time instant t_k to a centralized scheduler for decision making. Transmitting all state vectors, however, might induce considerable scheduling overhead. Under event-based control and scheduling, each plant is assigned a local event generator. Within each local event generator, an event-triggering law $\sigma_i(k) > 0$ with a relative-type threshold tuned by a design parameter $\lambda_i \in \mathbb{R}_0^+$ is implemented and checked at each time instant t_k. Based on the output of the event generators, the optimal switching index $j^*(k)$ is determined online based on an event-based switching law

$$j^*(k) = \begin{cases} 0 & \text{if } \sigma_i(k) \leq 0,\ \forall i \in \mathbb{M} \\ \arg\max\limits_{i\in\mathbb{M}} \sigma_i(k) & \text{otherwise.} \end{cases} \tag{1.21}$$

The proposed event based switching law (1.21) can be viewed as an extension of the time-based TOD protocol introduced in [WY01, WYB02], cf. Remark 7.3. It is worth to note that the proposed switching law does not require any state measurements for decision making, it is based on the *scalar* outputs of the event generators. Furthermore, it can be implemented in CAN-based communication networks in a decentralized fashion.

Based on the event-based switching law (1.21), Problem 1.1 is then transformed into a codesign problem of the event-triggering laws $\sigma_i(k)$ and the corresponding feedback matrices $\boldsymbol{K}_i$ for all $i \in \mathbb{M}$. The resulting codesign problem is formulated as an LMI optimization problem using the S-procedure, see Appendix A.3 for more details on the S-procedure. The effectiveness of the proposed strategy and the effect of the chosen tuning parameters λ_i, $\forall i \in \mathbb{M}$, on the resulting control performance are evaluated by simulation for an illustrative example and also experimentally for a case study.

Prediction-Based Control and Scheduling

This strategy can be seen as an extension of the event-based codesign strategy outlined above. Each event generator is equipped further with a local model-based predictor, leading to a model-based event generator. A replica of this predictor exists at the control node as well. Model-based event generators are quite known in event-based control, see e.g. [GA11, LLJ12, HD13]. They can significantly reduce the traffic load on the communication network due to the available information on system dynamics. In order to extend the event-based codesign strategy for model-based event generators, the switched system model (1.6) is first augmented with the predictor dynamics. Based on the augmented system model and the event-based switching law (1.21), the codesign problem of the model-based event-triggering laws $\sigma_i(k)$ and the corresponding feedback matrices $\boldsymbol{K}_i$ for all $i \in \mathbb{M}$ can now be formulated as an LMI optimization problem following the same line as for the event-based codesign problem. The effectiveness of the proposed strategy as well as the effect of the tuning parameters $\lambda_{\mathrm{x}i}$, $\forall i \in \mathbb{M}$, at the sensors side and λ_{u} at the controller side on the resulting control performance are evaluated by simulation for an illustrative example and also experimentally for a case study.

1.8 Outline and Publications

The thesis addresses the control and scheduling codesign problem for multiple control loops closed over a common communication network. Five methods to tackle the codesign problem are proposed. Several articles on the proposed methods have been published during the doctoral studies. The outline of the thesis along with references to the related articles are given below.

Chapter 2: Modeling. The architecture of the NCS used throughout the thesis is presented in Section 2.1. Based on this architecture, a mathematical model of the system dynamics including the network-induced imperfections is derived in Section 2.2. A polytopic overapproximation for the resulting model is formulated in Section 2.3, leading to the switched polytopic system with additive norm-bounded uncertainty (1.6). Finally, a quadratic cost function for evaluation purposes is defined in Section 2.4.

Chapter 3: Codesign Problem. The structure of the control law is first defined. Based on the NCS model derived in Chapter 2 and the control law, the main control and scheduling codesign problem is introduced in Section 3.1. The properties of the main codesign problem including convexity, tractability, and stability guarantee are discussed in Section 3.2.

Chapter 4: Periodic Control and Scheduling (PCS). The main codesign problem introduced in Chapter 3 is transformed for p-periodic switching sequences as a periodic control and scheduling codesign problem in Section 4.1. To get a tractable optimization problem, an upper bound on the objective function of the PCS codesign problem is then derived in Section 4.2. For each admissible p-periodic switching sequence, a p-periodic feedback sequence is designed in Section 4.3. Three solutions of the PCS codesign problem are proposed based on exhaustive search (Section 4.4), relaxation (Section 4.5), and optimal pointer placement (Section 4.6). The properties and effectiveness of the proposed PCS strategy are evaluated by an illustrative example in Section 4.7. This chapter is based on the following publication:

[AGL11] Sanad Al-Areqi, Daniel Görges, and Steven Liu. Robust control and scheduling codesign for networked embedded control systems. In *Proceedings of the 50th IEEE Conference on Decision and Control and European Control Conference 2011*, pages 3154–3159, 2011.

Chapter 5: Receding-Horizon Control and Scheduling (RHCS). A transformation of the main codesign problem into a receding-horizon control and scheduling codesign problem is outlined in Section 5.1. The RHCS problem is then solved based on dynamic programming in Section 5.2. A less complex solution using relaxed dynamic programming is presented in Section 5.3. The properties and effectiveness of the proposed RHCS strategy are evaluated by an illustrative example in Section 5.4. This chapter is based on the following publication:

[AGL12a] Sanad Al-Areqi, Daniel Görges, and Steven Liu. Receding-horizon control and scheduling of systems with uncertain computation and communication delays. In *Proceedings of the 51st IEEE Conference on Decision and Control*, pages 2654–2659, 2012.

Chapter 6: Implementation-Aware Control and Scheduling (IACS). The main codesign problem is transformed for a state-based switching law into a codesign problem of the scheduling parameters and the control parameters in Section 6.1. The implementation-aware control and scheduling codesign problem is then formulated as an LMI optimization problem and solved based on the Lyapunov-Metzler function in Section 6.2. Furthermore, feasibility of the LMI optimization problem and closed-loop stability are analyzed in Section 6.2.1 and Section 6.2.2, respectively. The properties and effectiveness of the proposed IACS strategy are evaluated by an illustrative example in Section 6.3. This chapter is based on the following publications:

[AGL12b] Sanad Al-Areqi, Daniel Görges, and Steven Liu. Robust feedback control and scheduling of networked embedded control systems. In *Proceedings of the 4th IFAC Conference on Analysis and Design of Hybrid Systems*, pages 127–132, 2012.

[RAL13a] Sven Reimann, Sanad Al-Areqi, and Steven Liu. An event-based online scheduling approach for networked embedded control systems. In *Proceedings of the American Control Conference*, pages 5326–5331, 2013.

[RAL13b] Sven Reimann, Sanad Al-Areqi, and Steven Liu. Output-based control and scheduling codesign for control systems sharing a limited resource. In *Proceedings of the 52nd IEEE Conference on Decision and Control*, pages 4042–4047, 2013.

Chapter 7: Event-Based Control and Scheduling (EBCS). The architecture of the NCS introduced in Chapter 2 is extended in Section 7.1 such that a local event generator is assigned to each plant. A mathematical model of the system dynamics based on the extended architecture is derived as well. Based on the resulting NCS model, the event-based control and scheduling codesign problem is introduced in Section 7.2. The solution of the EBCS codesign problem based on the S-procedure is detailed in Section 7.3. The properties and effectiveness of the proposed EBCS strategy are evaluated by an illustrative example in Section 7.4. This chapter is based on the following publications:

[AGL13a] Sanad Al-Areqi, Daniel Görges, and Steven Liu. Event-based control and scheduling codesign of networked embedded control systems. In *Proceedings of the American Control Conference*, pages 5299–5304, 2013.

[AGL13b] Sanad Al-Areqi, Daniel Görges, and Steven Liu. Robust event-based control and scheduling of networked embedded control systems. In *Proceedings of the 4th IFAC Workshop on Distributed Estimation and Control in Networked Systems*, pages 7–14, 2013.

[AGL14] Sanad Al-Areqi, Daniel Görges, and Steven Liu. Stochastic event-based control and scheduling of large-scale networked control systems. In *Proceedings of the European Control Conference*, pages 2316–2321, 2014.

[AGL15c] Sanad Al-Areqi, Daniel Görges, and Steven Liu. Event-based networked control and scheduling codesign with guaranteed performance. *Automatica*, 57:128–134, 2015.

[AGL15a] Sanad Al-Areqi, Daniel Görges, and Steven Liu. Event-based control and scheduling codesign: Stochastic and robust approaches. *IEEE Transactions on Automatic Control*, 60(5):1291–1303, 2015.

[RVAL15] Sven Reimann, Duc Hai Van, Sanad Al-Areqi, and Steven Liu. Stability Analysis and PI Control Synthesis under Event-Triggered Communication. In *Proceedings of the 2015 European Control Conference*, pages 2179–2184, 2015.

Chapter 8: Prediction-Based Control and Scheduling (PBCS). The architecture of the NCS introduced in Chapter 2 is further extended in Section 8.1 such that each local event generator is equipped with a model-based predictor. An augmented NCS model including the system dynamics and the predictor dynamics is derived as well. Based on the resulting augmented NCS model, the prediction-based control and scheduling codesign problem is introduced in Section 8.2. The solution of the PBCS codesign problem based on the S-procedure is detailed in Section 8.3. The properties and effectiveness of the proposed PBCS strategy are evaluated by an illustrative example in Section 8.4. This chapter is based on the following publications:

[AGL15b] Sanad Al-Areqi, Daniel Görges, and Steven Liu. Event-Based Control and Scheduling Codesign Subject to Input and State Constraints. In *Proceedings of the European Control Conference*, pages 1860–1865, 2015.

[WARL15] Benjamin Watkins, Sanad Al-Areqi, Sven Reimann, and Steven Liu. Event-Based Control of Constrained Discrete-Time Linear Systems with Guaranteed Performance. *International Journal of Sensors, Wireless Communications, and Control*, 5(2):72–80, 2015.

Chapter 9: Evaluation and Implementation. The proposed control and scheduling codesign strategies proposed in the thesis are evaluated and compared simulatively in Section 9.1 and experimentally in Section 9.2. The simulation setup is detailed in Section 9.1.1. Based on the simulation setup, the results are illustrated in Section 9.1.2 and conclusions are drawn in Section 9.1.3. For the experimental study, the setup is outlined in Section 9.2.1. Based on the setup, the experimental results of each codesign strategy are given in Section 9.2.2. Conclusions are finally provided in Section 9.2.3.

Chapter 10: Conclusions and Future Work. Conclusions on the content of the thesis are drawn in Section 10.1. Furthermore, recommendations and suggestions for future work are provided in Section 10.2.

2 Modeling

In this chapter, the architecture of the NCS under consideration is first introduced. Based on this architecture, a formal model of system dynamics is then derived while taking the network-induced imperfections outlined in Section 1.3 explicitly into account. A quadratic cost function is further assigned to the NCS for the purpose of formulating the codesign problem ,introduced later, as an optimization problem and also for performance evaluation. The resulting system model together with the cost function are used in the following chapters, with minor changes if necessary, for control and scheduling codesign purposes.

2.1 NCS Architecture

Consider the NCS with a set of *individual* control loops as illustrated in Figure 2.1.

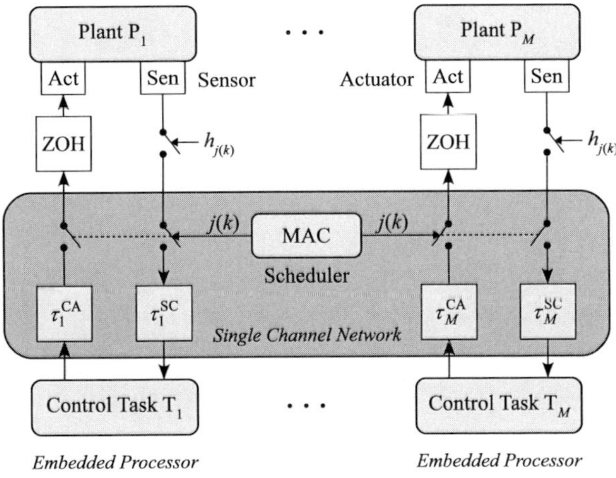

Figure 2.1: Architecture of a single channel networked control system

Each control loop consists of a plant P_i associated with a sensor, an actuator, and a control task T_i implemented on an embedded processor $\forall i \in \mathbb{M} = \{1, \ldots, M\}$. The control loops are closed over a shared communication network. The access to the network is scheduled by a medium access control (MAC) unit. Besides the medium access constraints, each control loop experiences uncertain time-varying delays $\tau_i^{\mathrm{SC}}(k) \in [\underline{\tau}_i^{\mathrm{SC}}, \overline{\tau}_i^{\mathrm{SC}}]$ required for transmitting new measurements to the controller and $\tau_i^{\mathrm{CA}}(k) \in [\underline{\tau}_i^{\mathrm{CA}}, \overline{\tau}_i^{\mathrm{CA}}]$ required for transmitting new control updates to the corresponding actuators. The lower bounds $\underline{\tau}_i^{\mathrm{SC}}$, $\underline{\tau}_i^{\mathrm{CA}}$ and the upper bounds $\overline{\tau}_i^{\mathrm{SC}}$, $\overline{\tau}_i^{\mathrm{CA}}$ for all $i \in \mathbb{M}$ are assumed to be known. Finally, due to the digital nature of the implementation platform (the communication network and the embedded processors) each control loop is equipped with a zero-order hold (ZOH) device and a sampling device. The sampling interval $h_{j(k)}$ is time-varying due to its dependency on the output of the scheduler $j(k) \in \mathbb{M}$, but belongs to a finite set. Further assumptions on $h_{j(k)}$ are given in the following section.

Assumption 2.1 *It is assumed within this thesis that the utilized communication network is synchronized. This assumption is required for a centralized coordination (scheduling) of the transmissions over the network, avoiding any collision and thus improving the control performance. Synchronous behavior can be provided in several ways, including external clock sources such as GPS or over-the-air clock synchronization and internal message exchange synchronization protocols. Various synchronization protocols for wired and wireless networks have been developed over the past few decades. Most of them are characterized by high synchronization precision, i.e. maximum deviation (skew and offset) among distributed clocks is in the range of microseconds, see [SBK05] for further details. Typical examples of synchronous networks include time-triggered protocol (TTP), time-triggered CAN (TT-CAN), and FlexRay [NHB05].*

For notational convenience, the index $k \in \mathbb{N}_0$ is used in the thesis for representing a sampling instant t_k if no ambiguity arises.

2.2 NCS Model

Each plant P_i together with its associated sensors and actuators is described by a continuous-time LTI state equation

$$\begin{aligned} \dot{\boldsymbol{x}}_{\mathrm{c}i}(t) &= \boldsymbol{A}_{\mathrm{c}i}\boldsymbol{x}_{\mathrm{c}i}(t) + \boldsymbol{B}_{\mathrm{c}i}\boldsymbol{u}_i(t) \\ \boldsymbol{x}_{\mathrm{c}i}(0) &= \boldsymbol{x}_{\mathrm{c}i0} \end{aligned} \tag{2.1}$$

where $\boldsymbol{A}_{\mathrm{c}i} \in \mathbb{R}^{n_i \times n_i}$ is the system matrix, $\boldsymbol{B}_{\mathrm{c}i} \in \mathbb{R}^{n_i \times m_i}$ is the input matrix, $\boldsymbol{x}_{\mathrm{c}i}(t) \in \mathbb{R}^{n_i}$ is the state vector and $\boldsymbol{u}_i(t) \in \mathbb{R}^{m_i}$ is the control vector. For the purpose of integrating the network-induced imperfections such as medium access constraints and variable transmission times into the state equation (2.1), a discrete-time model of (2.1) is considered. The continuous-time state equation (2.1) is discretized over the sampling interval

$t_k \leq t < t_{k+1}$ using ZOH. Because of the medium access constraints with which only one plant gets access to the communication network within the sampling interval, two cases must be distinguished:

- *Open-loop* plant P_i, where the output of the scheduler in Figure 2.1 is $j(k) \neq i$. Thus, no new measurements are sent for executing the control task T_i. Consequently, the last sent control vector $\hat{\boldsymbol{u}}_i(t_k)$ is further held by the ZOH, i.e.

$$\boldsymbol{u}_i(t) = \hat{\boldsymbol{u}}_i(t_k) \quad \text{for} \quad t_k \leq t < t_{k+1}. \tag{2.2}$$

- *Closed-loop* plant P_i, where the output of the scheduler is $j(k) = i$. In this case, new measurements are sent for executing the control task T_i. The resulting control update is sent further to the corresponding actuator, leading to

$$\boldsymbol{u}_i(t) = \begin{cases} \hat{\boldsymbol{u}}_i(t_k) & \text{for} \quad t_k \leq t < t_k + \tau_i(k) \\ \boldsymbol{u}_i(t_k) & \text{for} \quad t_k + \tau_i(k) \leq t < t_{k+1} \end{cases} \tag{2.3}$$

with the input delay $\tau_i(k) \in [\underline{\tau}_i, \overline{\tau}_i]$ subsuming the network-induced delays within the control loop, i.e.

$$\tau_i(k) = \tau_i^{\mathrm{SC}}(k) + \tau_i^{\mathrm{CA}}(k). \tag{2.4}$$

Assumption 2.2 *The induced input delay $\tau_i(k)$ is assumed to be not longer than the chosen sampling interval $h_i = t_{k+1} - t_k$ for all $i \in \mathbb{M}$. This technical assumption is made for the purpose of simplifying the presentation. The case where $\tau_i(k)$ is longer than h_i is discussed in Remark 2.6 in more detail.*

Remark 2.1. The sampling interval $h_{j(k)}$ chosen at time instant t_k depends on the output of the scheduler $j(k)$. A question may arise here about how to inform the samplers located close to the sensors about the next sampling instant t_{k+1}. In fact, in single-channel networks all connected nodes can listen simultaneously to the channel and thus be informed about the actual output of the scheduler. Based on the output of the scheduler and Assumption 2.1, the next sampling instant t_{k+1} can be determined provided that $h_{j(k)}$, $\forall j(k) \in \mathbb{M}$, is known to all nodes.

The distinction above is graphically illustrated in Figure 2.2 and included in the resulting discrete-time model using the logical variable

$$\delta_{ij(k)} = \begin{cases} 1 & \text{if} \quad i = j(k) \\ 0 & \text{if} \quad i \neq j(k)\,. \end{cases} \tag{2.5}$$

Considering this distinction, the solution of the continuous-time state equation (2.1) at the sampling instants t_k, $k \in \mathbb{N}$, is given by

$$\boldsymbol{x}_{\mathrm{c}i}(t_{k+1}) = \boldsymbol{\Phi}_i(h_{j(k)})\boldsymbol{x}_{\mathrm{c}i}(t_k) + \boldsymbol{\Gamma}_{1i}(h_{j(k)}, \tau_{ij(k)})\hat{\boldsymbol{u}}_i(t_k) + \boldsymbol{\Gamma}_{0i}(h_{j(k)}, \tau_{ij(k)})\boldsymbol{u}_i(t_k) \tag{2.6}$$

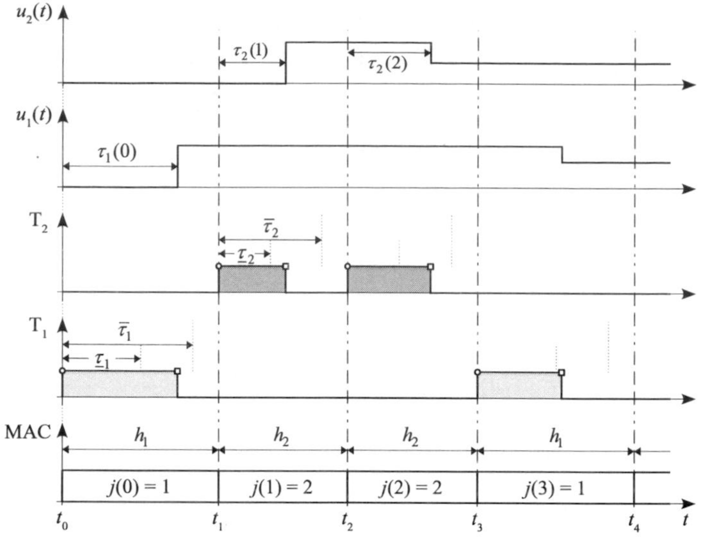

Figure 2.2: NECS timing diagram. A new measurement is indicated by a circle and a control update by a square

where

$$\boldsymbol{\Phi}_i(h_{j(k)}) = e^{\boldsymbol{A}_{ci} h_{j(k)}} \tag{2.7a}$$
$$\boldsymbol{\Gamma}_{1i}(h_{j(k)}, \tau_{ij(k)}) = \boldsymbol{\Gamma}_i(h_{j(k)}) - \boldsymbol{\Gamma}_i(h_{j(k)} - \tau_{ij(k)}) \tag{2.7b}$$
$$\boldsymbol{\Gamma}_{0i}(h_{j(k)}, \tau_{ij(k)}) = \boldsymbol{\Gamma}_i(h_{j(k)} - \tau_{ij(k)}) \tag{2.7c}$$
$$\boldsymbol{\Gamma}_i(t) = \int_0^t e^{\boldsymbol{A}_{ci} s} ds \boldsymbol{B}_{ci} \tag{2.7d}$$
$$\tau_{ij(k)} = \delta_{ij(k)} \tau_i(k) + (1 - \delta_{ij(k)}) h_{j(k)}. \tag{2.7e}$$

Introducing the augmented state vector $\boldsymbol{x}_i(k) = \begin{pmatrix} \boldsymbol{x}_{ci}^T(k) & \hat{\boldsymbol{u}}_i^T(k) \end{pmatrix}^T$, the resulting discrete-time state equation corresponding to (2.1) is given by

$$\begin{aligned} \boldsymbol{x}_i(k+1) &= \boldsymbol{A}_{ij(k)}(k) \boldsymbol{x}_i(k) + \boldsymbol{B}_{ij(k)}(k) \boldsymbol{u}_i(k) \\ \boldsymbol{x}_i(0) &= \begin{pmatrix} \boldsymbol{x}_{ci0}^T & \boldsymbol{0} \end{pmatrix}^T \end{aligned} \tag{2.8}$$

where $\boldsymbol{A}_{ij(k)}(k) \in \mathbb{R}^{(n_i+m_i)\times(n_i+m_i)}$ and $\boldsymbol{B}_{ij(k)}(k) \in \mathbb{R}^{(n_i+m_i)\times m_i}$ are constructed as

$$\boldsymbol{A}_{ij(k)}(k) = \begin{pmatrix} \boldsymbol{\Phi}_i(h_{j(k)}) & \boldsymbol{\Gamma}_i(h_{j(k)}) - \boldsymbol{\Gamma}_i(h_{j(k)} - \tau_{ij(k)}) \\ \boldsymbol{0} & (1-\delta_{ij(k)})\boldsymbol{I} \end{pmatrix} \tag{2.9a}$$

$$\boldsymbol{B}_{ij(k)}(k) = \begin{pmatrix} \boldsymbol{\Gamma}_i(h_{j(k)} - \tau_{ij(k)}) \\ \delta_{ij(k)}\boldsymbol{I} \end{pmatrix} . \tag{2.9b}$$

Remark 2.2. The discrete-time state equation (2.8), adopted from a modeling of time-delay systems proposed in [ÅW97, Sec. 2.3], describes the dynamics of plant P_i taking the uncertain input delay $\tau_i(k)$ and the medium access constraints into account.

The NCS shown in Figure 2.1 can now be represented by a block-diagonal discrete-time switched linear system

$$\begin{aligned} \boldsymbol{x}(k+1) &= \boldsymbol{A}_{j(k)}(k)\boldsymbol{x}(k) + \boldsymbol{B}_{j(k)}(k)\boldsymbol{u}(k) \\ \boldsymbol{x}(0) &= \begin{pmatrix} \boldsymbol{x}_1^T(0) & \cdots & \boldsymbol{x}_M^T(0) \end{pmatrix}^T \end{aligned} \tag{2.10}$$

where

$$\boldsymbol{x}(k) = \begin{pmatrix} \boldsymbol{x}_1^T(k) & \cdots & \boldsymbol{x}_M^T(k) \end{pmatrix}^T \in \mathbb{R}^{n+m} \tag{2.11a}$$

$$\boldsymbol{u}(k) = \begin{pmatrix} \boldsymbol{u}_1^T(k) & \cdots & \boldsymbol{u}_M^T(k) \end{pmatrix}^T \in \mathbb{R}^{m} \tag{2.11b}$$

$$\boldsymbol{A}_{j(k)}(k) = \operatorname{diag}\left(\boldsymbol{A}_{1j(k)}(k), \ldots, \boldsymbol{A}_{Mj(k)}(k)\right) \tag{2.11c}$$

$$\boldsymbol{B}_{j(k)}(k) = \operatorname{diag}\left(\boldsymbol{B}_{1j(k)}(k), \ldots, \boldsymbol{B}_{Mj(k)}(k)\right) . \tag{2.11d}$$

The dimensions n and m are given by

$$n = \sum_{i=1}^{M} n_i, \qquad m = \sum_{i=1}^{M} m_i. \tag{2.12}$$

The resulting NCS model (2.10) describes exactly the dynamics of all plants at time instants t_k taking the network-induced imperfections, shown in Figure 2.1, into account.

Remark 2.3. Note that a similar model has been derived in [GIL09] for networked control systems with constant input delays, i.e. $\tau_i(k) = \tau_i$, $\forall k \in \mathbb{N}_0$. This model is generalized here for dealing with uncertain time-varying input delays.

Remark 2.4. In case of the plants P_i, $\forall i \in \mathbb{M}$, in Figure 2.1 are not independent of each other but interacted, the modeling procedure proposed here can be extended in a straightforward manner as shown in [AGL13b].

Note that the system matrix $\boldsymbol{A}_{ij(k)}(k)$ and the input matrix $\boldsymbol{B}_{ij(k)}(k)$ in case of $\delta_{ij(k)} = 1$ are affected by the uncertain time-varying input delay $\tau_i(k) \in [\underline{\tau}_i, \overline{\tau}_i]$ via the matrix $\boldsymbol{\Gamma}_i(h_{j(k)} - \tau_{ij(k)}) \in \boldsymbol{\Omega}_i$ with

$$\boldsymbol{\Omega}_i \triangleq \left\{ \boldsymbol{\Gamma}_i(h_{j(k)} - \tau_{ij(k)}) \mid \tau_i(k) \in [\underline{\tau}_i, \overline{\tau}_i],\, \delta_{ij(k)} = 1,\, \forall k \in \mathbb{N}_0 \right\} . \tag{2.13}$$

Because of the exponential dependency of $\boldsymbol{\Gamma}_i(h_{j(k)} - \tau_{ij(k)})$ on $\tau_i(k)$, seen from the equation (2.7d), the set of matrices $\boldsymbol{\Omega}_i$ in (2.13) can even be non-convex. Such structural properties of the uncertain time-varying discrete-time state equation (2.8) make any analysis of the closed-loop stability or a synthesis of stabilizing controller a hard problem. An overapproximation of the possibly non-convex set of matrices $\boldsymbol{\Omega}_i$ by a convex polytopic set is therefore employed in the following section. By doing this, the analysis and design tools proposed in robust control theory [KBM96, DB01] for linear systems with polytopic uncertainty can be applied as shown in the following chapters.

2.3 Polytopic Formulation

In this section, we aim at overapproximating the possibly non-convex set of matrices $\boldsymbol{\Omega}_i$ in (2.13) by a convex polytopic set. Various methods have been proposed during the last years to obtain a convex polytopic overapproximation with (or without) additive norm-bounded uncertainties, e.g. based on gridding [BK08,SB09,DHWH11], the Jordan normal form [CWHN09], Taylor series expansion [HDL06] or the Cayley-Hamilton Theorem [GOL+10]. A recent survey with an assessment of the complexity and conservatism is given in [HWG+10]. In the following, the Taylor series expansion method is used for extracting the convex polytopic set with additive norm-bounded uncertainty. Noteworthy, the overapproximation is only required for $\delta_{ij(k)} = 1$, i.e. when $j(k) = i$. Otherwise, no overapproximation is required since the plant model (2.8) does not depend in this case on the uncertain parameter $\tau_i(k)$, the only source of uncertainty.

The first step of extracting the convex polytopic set using the Taylor series method consists in expanding the matrix exponential contained in the matrix

$$\boldsymbol{\Gamma}_i\big(h_{j(k)} - \tau_{ij(k)}\big) = \boldsymbol{\Gamma}_{j(k)}\big(h_{j(k)} - \tau_{j(k)}(k)\big) \quad \text{for} \quad \delta_{ij(k)} = 1. \tag{2.14}$$

The resulting Taylor series are then partitioned into an approximation part and a remainder part

$$\begin{aligned}
\boldsymbol{\Gamma}_i(\rho_{j(k)}) &= \int_0^{\rho_{j(k)}} \sum_{q=0}^{\infty} \frac{\boldsymbol{A}_{\text{ci}}^q}{q!} s^q ds \boldsymbol{B}_{\text{ci}} \overset{\ell=q+1}{=} \sum_{\ell=1}^{\infty} \frac{\boldsymbol{A}_{\text{ci}}^{\ell-1}}{\ell!} \rho_{j(k)}^{\ell} \boldsymbol{B}_{\text{ci}} && \text{(2.15a)}\\
&= \sum_{\ell=1}^{L} \frac{\boldsymbol{A}_{\text{ci}}^{\ell-1}}{\ell!} \rho_{j(k)}^{\ell} \boldsymbol{B}_{\text{ci}} + \sum_{\ell=L+1}^{\infty} \frac{\boldsymbol{A}_{\text{ci}}^{\ell-1}}{\ell!} \rho_{j(k)}^{\ell} \boldsymbol{B}_{\text{ci}} && \text{(2.15b)}\\
&= \hat{\boldsymbol{\Gamma}}_i(\rho_{j(k)}, L) + \Delta\boldsymbol{\Gamma}_i(\rho_{j(k)}, L). && \text{(2.15c)}
\end{aligned}$$

with the uncertain parameter $\rho_{j(k)} = h_{j(k)} - \tau_{j(k)}(k)$. Partitioning the Taylor series in an approximation part $\hat{\boldsymbol{\Gamma}}_i(\rho_{j(k)}, L)$ (matrix polynomial of order L) and a remainder $\Delta\boldsymbol{\Gamma}_i(\rho_{j(k)}, L)$ allows the polytopic formulation of the approximation part. The order L of the matrix polynomial $\hat{\boldsymbol{\Gamma}}_i(\rho_{j(k)}, L)$ is a tuning parameter left to the designer. The higher the chosen order L, the lower the resulting additive uncertainty, however, at the cost of more complexity in the codesign later on. The matrix polynomial $\hat{\boldsymbol{\Gamma}}_i(\rho_{j(k)}, L)$ can now be enveloped by a convex polytope according to Lemma A.1, leading finally to

$$\hat{\boldsymbol{\Gamma}}_i(\rho_{j(k)}, L) = \sum_{\ell=1}^{L+1} \mu_\ell(\rho_{j(k)}) \hat{\boldsymbol{\Gamma}}_{i\ell}(\underline{\rho}_{j(k)}, \overline{\rho}_{j(k)}) \tag{2.16}$$

with the non-negative real scalars $\mu_\ell(\rho_{j(k)}) \in \mathbb{R}_0^+$ such that $\sum_{\ell=1}^{L+1} \mu_\ell(\rho_{j(k)}) = 1$ and the constant polytope vertices $\hat{\boldsymbol{\Gamma}}_{i\ell}(\underline{\rho}_{j(k)}, \overline{\rho}_{j(k)})$. The overapproximation procedure based on the Taylor series method is graphically illustrated in Figure 2.3.

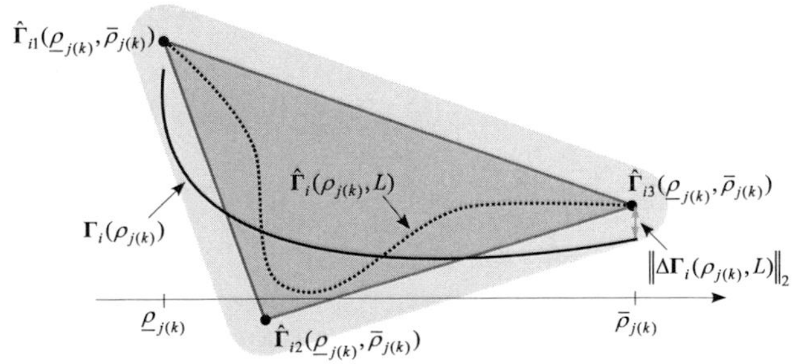

Figure 2.3: An overapproximation of $\boldsymbol{\Gamma}_i(\rho_{j(k)})$ based on the Taylor series method

Substituting (2.16) into (2.15c) and further (2.15c) into (2.9) results in, after factorizing $\mu_\ell(\rho_{j(k)})$, a discrete-time plant model with polytopic and additive norm-bounded uncertainty

$$\begin{aligned} \boldsymbol{x}_i(k+1) = & \left(\sum_{\ell=1}^{L+1} \mu_\ell(\rho_{j(k)}) \boldsymbol{A}_{ij(k)\ell} + \boldsymbol{D}_{ij(k)} \boldsymbol{E}_{ij(k)} \boldsymbol{F}_{\mathrm{ai}} \right) \boldsymbol{x}_i(k) \\ & + \left(\sum_{\ell=1}^{L+1} \mu_\ell(\rho_{j(k)}) \boldsymbol{B}_{ij(k)\ell} + \boldsymbol{D}_{ij(k)} \boldsymbol{E}_{ij(k)} \boldsymbol{F}_{\mathrm{bi}} \right) \boldsymbol{u}_i(k) \end{aligned} \tag{2.17}$$

where $\boldsymbol{A}_{ij(k)\ell} \in \mathbb{R}^{(n_i+m_i)\times(n_i+m_i)}$, $\boldsymbol{B}_{ij(k)\ell} \in \mathbb{R}^{(n_i+m_i)\times m_i}$, $\boldsymbol{D}_{ij(k)} \in \mathbb{R}^{(n_i+m_i)\times n_i}$, $\boldsymbol{E}_{ij(k)} \in \mathbb{R}^{n_i \times m_i}$, $\boldsymbol{F}_{\mathrm{ai}} \in \mathbb{R}^{m_i\times(n_i+m_i)}$, and $\boldsymbol{F}_{\mathrm{bi}} \in \mathbb{R}^{m_i \times m_i}$ are constructed as

$$\begin{aligned} \boldsymbol{A}_{ij(k)\ell} &= \begin{pmatrix} \boldsymbol{\Phi}_i(h_{j(k)}) & \boldsymbol{\Gamma}_i(h_{j(k)}) - \hat{\boldsymbol{\Gamma}}_{i\ell}(\underline{\rho}_{j(k)}, \overline{\rho}_{j(k)}) \\ \boldsymbol{0} & (1-\delta_{ij(k)})\boldsymbol{I} \end{pmatrix}, & \boldsymbol{E}_{ij(k)} &= \gamma_{ij(k)}^{-1} \Delta\boldsymbol{\Gamma}_i(\rho_{j(k)}, L) \\ \boldsymbol{B}_{ij(k)\ell} &= \begin{pmatrix} \hat{\boldsymbol{\Gamma}}_{i\ell}(\underline{\rho}_{j(k)}, \overline{\rho}_{j(k)}) \\ \delta_{ij(k)}\boldsymbol{I} \end{pmatrix}, & \boldsymbol{F}_{\mathrm{ai}} &= \begin{pmatrix} \boldsymbol{0} & -\boldsymbol{I} \end{pmatrix} \\ \boldsymbol{D}_{ij(k)} &= \begin{pmatrix} \gamma_{ij(k)}\boldsymbol{I} \\ \boldsymbol{0} \end{pmatrix}, & \boldsymbol{F}_{\mathrm{bi}} &= \boldsymbol{I} \end{aligned}$$

with $\|\Delta\boldsymbol{\Gamma}_i(\rho_{j(k)}, L)\|_2 \le \gamma_{ij(k)}$ ensuring $\|\boldsymbol{E}_{ij(k)}\|_2 \le 1$.

Remark 2.5. The factor $\gamma_{ij(k)} \in \mathbb{R}_0^+$ can be determined numerically by gridding the uncertain parameter $\rho_{j(k)} \in [\underline{\rho}_{j(k)}, \overline{\rho}_{j(k)}]$ and computing the induced 2-norm $\left\|\Delta\boldsymbol{\Gamma}_i(\rho_{j(k)}, L)\right\|_2$ for each grid point. Alternatively, the remainder $\Delta\boldsymbol{\Gamma}_i(\rho_{j(k)}, L)$ defined in (2.15c) can first be overapproximated using the Jordan normal form or the Cayley-Hamilton theorem. Based on the resulting polytope, the condition $\left\|\Delta\boldsymbol{\Gamma}_i(\rho_{j(k)}, L)\right\|_2 \leq \gamma_{ij(k)}$ is then formulated as an LMI optimization problem with the objective of determining the minimum value of $\gamma_{ij(k)}$.

Substituting finally the plant model (2.17) into the NCS model (2.10) yields a block-diagonal discrete-time switched polytopic system with additive norm-bounded uncertainty

$$\begin{aligned}\boldsymbol{x}(k+1) = &\left(\sum_{\ell=1}^{L+1} \mu_\ell(\rho_{j(k)})\boldsymbol{A}_{j(k)\ell} + \boldsymbol{D}_{j(k)}\boldsymbol{E}_{j(k)}\boldsymbol{F}_{\mathrm{a}}\right)\boldsymbol{x}(k) \\ &+ \left(\sum_{\ell=1}^{L+1} \mu_\ell(\rho_{j(k)})\boldsymbol{B}_{j(k)\ell} + \boldsymbol{D}_{j(k)}\boldsymbol{E}_{j(k)}\boldsymbol{F}_{\mathrm{b}}\right)\boldsymbol{u}(k)\end{aligned} \tag{2.18}$$

where the vectors $\boldsymbol{x}(k)$, $\boldsymbol{u}(k)$ are defined in (2.11) and the matrices $\boldsymbol{A}_{j(k)\ell} \in \mathbb{R}^{(n+m)\times(n+m)}$, $\boldsymbol{B}_{j(k)\ell} \in \mathbb{R}^{(n+m)\times m}$, $\boldsymbol{D}_{j(k)} \in \mathbb{R}^{(n+m)\times n}$, $\boldsymbol{E}_{j(k)} \in \mathbb{R}^{n\times m}$, $\boldsymbol{F}_{\mathrm{a}} \in \mathbb{R}^{m\times(n+m)}$, and $\boldsymbol{F}_{\mathrm{b}} \in \mathbb{R}^{m\times m}$ are constructed as

$$\boldsymbol{A}_{j(k)\ell} = \mathrm{diag}\left(\boldsymbol{A}_{1j(k)\ell}, \ldots, \boldsymbol{A}_{Mj(k)\ell}\right) \tag{2.19a}$$
$$\boldsymbol{B}_{j(k)\ell} = \mathrm{diag}\left(\boldsymbol{B}_{1j(k)\ell}, \ldots, \boldsymbol{B}_{Mj(k)\ell}\right) \tag{2.19b}$$
$$\boldsymbol{D}_{j(k)} = \mathrm{diag}\left(\boldsymbol{D}_{1j(k)}, \ldots, \boldsymbol{D}_{Mj(k)}\right) \tag{2.19c}$$
$$\boldsymbol{E}_{j(k)} = \mathrm{diag}\left(\boldsymbol{E}_{1j(k)}, \ldots, \boldsymbol{E}_{Mj(k)}\right) \tag{2.19d}$$
$$\boldsymbol{F}_{\mathrm{a}} = \mathrm{diag}\left(\boldsymbol{F}_{\mathrm{a}1}, \ldots, \boldsymbol{F}_{\mathrm{a}M}\right) \tag{2.19e}$$
$$\boldsymbol{F}_{\mathrm{b}} = \mathrm{diag}\left(\boldsymbol{F}_{\mathrm{b}1}, \ldots, \boldsymbol{F}_{\mathrm{b}M}\right). \tag{2.19f}$$

For brevity, the discrete-time switched polytopic system with additive norm-bounded uncertainty (2.18) is denoted in the following as switched polytopic system if no ambiguity arises.

Remark 2.6. Two network-induced imperfections have been explicitly considered during the modeling procedure of the NCS, namely variable transmission times and medium access constraints. The variable transmission time has been represented as an uncertain time-varying input delay $\tau_{j(k)}(k)$ which is not longer than the chosen sampling interval $h_{j(k)}$. In case of $\tau_{j(k)}(k)$ longer than $h_{j(k)}$, a minor change of the proposed system modeling is then required following any of the ideas proposed in [Het07] or in [CWHN09].

Remark 2.7. In case of uncertain (rather than certain) time-varying sampling intervals, the system modeling proposed above can be extended in a straightforward manner as discussed in [IGL10, DHWH11]. Thus, the variable sampling instants imperfection as well as the packet dropouts imperfection represented implicitly as a prolongation of sampling intervals can also be accounted for.

2.4 Cost Function

For the purpose of formulating the control and scheduling codesign problem as an optimization problem such that a certain criterion is minimized, each control loop is assigned a continuous-time quadratic cost function

$$J_i = \int_0^{\infty} \boldsymbol{x}_{\mathrm{c}i}^T(t)\boldsymbol{Q}_{\mathrm{c}i}\boldsymbol{x}_{\mathrm{c}i}(t) + \boldsymbol{u}_i^T(t)\boldsymbol{R}_{\mathrm{c}i}\boldsymbol{u}_i(t)dt \tag{2.20}$$

with the state weighting matrix $\boldsymbol{Q}_{ci} \in \mathbb{R}^{n_i \times n_i}$ and the input weighting matrix $\boldsymbol{R}_{\mathrm{c}i} \in \mathbb{R}^{m_i \times m_i}$ being symmetric and positive definite. The continuous-time quadratic cost function (2.20) can equivalently be rewritten as

$$J_i = \sum_{k=0}^{\infty} \int_{t_k}^{t_{k+1}} \boldsymbol{x}_{\mathrm{c}i}^T(t)\boldsymbol{Q}_{\mathrm{c}i}\boldsymbol{x}_{\mathrm{c}i}(t) + \boldsymbol{u}_i^T(t)\boldsymbol{R}_{\mathrm{c}i}\boldsymbol{u}_i(t)dt. \tag{2.21}$$

The evolution of the state $\boldsymbol{x}_{\mathrm{c}i}(t)$ over the sampling interval $t_k \leq t < t_{k+1}$ follows from the solution of the state equation (2.1) for a given state vector $\boldsymbol{x}_{\mathrm{c}i}(t_k)$, i.e.

$$\boldsymbol{x}_{\mathrm{c}i}(t) = e^{\boldsymbol{A}_{\mathrm{c}i}(t-t_k)}\boldsymbol{x}_{\mathrm{c}i}(t_k) + \int_{t_k}^{t} e^{\boldsymbol{A}_{\mathrm{c}i}(t-s)}\boldsymbol{B}_{\mathrm{c}i}\boldsymbol{u}_i(s)ds \tag{2.22}$$

Substituting (2.22) into (2.21) yields

$$J_i = \sum_{k=0}^{\infty} [I_1(k) + I_2(k) + I_3(k) + I_4(k) + I_5(k)] \tag{2.23}$$

where

$$I_1(k) = \int_{t_k}^{t_{k+1}} \left[e^{\boldsymbol{A}_{\mathrm{c}i}(t-t_k)}\boldsymbol{x}_{\mathrm{c}i}(t_k)\right]^T \boldsymbol{Q}_{\mathrm{c}i} \left[e^{\boldsymbol{A}_{\mathrm{c}i}(t-t_k)}\boldsymbol{x}_{\mathrm{c}i}(t_k)\right] dt \tag{2.24a}$$

$$I_2(k) = \int_{t_k}^{t_{k+1}} \left[e^{\boldsymbol{A}_{\mathrm{c}i}(t-t_k)}\boldsymbol{x}_{\mathrm{c}i}(t_k)\right]^T \boldsymbol{Q}_{\mathrm{c}i} \left[\int_{t_k}^{t} e^{\boldsymbol{A}_{\mathrm{c}i}(t-s)}\boldsymbol{B}_{\mathrm{c}i}\boldsymbol{u}_i(s)ds\right] dt \tag{2.24b}$$

$$I_3(k) = \int_{t_k}^{t_{k+1}} \left[\int_{t_k}^{t} e^{\boldsymbol{A}_{\mathrm{c}i}(t-s)}\boldsymbol{B}_{\mathrm{c}i}\boldsymbol{u}_i(s)ds\right]^T \boldsymbol{Q}_{\mathrm{c}i} \left[e^{\boldsymbol{A}_{\mathrm{c}i}(t-t_k)}\boldsymbol{x}_{\mathrm{c}i}(t_k)\right] dt \tag{2.24c}$$

$$I_4(k) = \int_{t_k}^{t_{k+1}} \left[\int_{t_k}^{t} e^{\boldsymbol{A}_{\mathrm{c}i}(t-s)}\boldsymbol{B}_{\mathrm{c}i}\boldsymbol{u}_i(s)ds\right]^T \boldsymbol{Q}_{\mathrm{c}i} \left[\int_{t_k}^{t} e^{\boldsymbol{A}_{\mathrm{c}i}(t-s)}\boldsymbol{B}_{\mathrm{c}i}\boldsymbol{u}_i(s)ds\right] dt \tag{2.24d}$$

$$I_5(k) = \int_{t_k}^{t_{k+1}} \boldsymbol{u}_i^T(t)\boldsymbol{R}_{\mathrm{c}i}\boldsymbol{u}_i(t)dt. \tag{2.24e}$$

Splitting the integrals $\int_{t_k}^{t_{k+1}}(\cdot)dt$ in (2.24) into two parts $\int_{t_k}^{t_k+\tau_{ij(k)}}(\cdot)dt + \int_{t_k+\tau_{ij(k)}}^{t_{k+1}}(\cdot)dt$ allows the substitution of the piecewise constant control vector

$$\boldsymbol{u}_i(t) = \begin{cases} \hat{\boldsymbol{u}}_i(t_k) & \text{for} \quad t_k \leq t < t_k + \tau_{ij(k)} \\ \boldsymbol{u}_i(t_k) & \text{for} \quad t_k + \tau_{ij(k)} \leq t < t_{k+1} \end{cases} \tag{2.25}$$

where $\tau_{ij(k)}$ is defined according to (2.7e) and the equation (2.25) combines both (2.2) and (2.3). The resulting discrete-time cost function, after factoring out $\boldsymbol{x}_{ci}(k)$, $\hat{\boldsymbol{u}}_i(k)$ and $\boldsymbol{u}_i(k)$, is then given by

$$J_i = \sum_{k=0}^{\infty} \begin{pmatrix} \boldsymbol{x}_{ci}(k) \\ \hat{\boldsymbol{u}}_i(k) \\ \boldsymbol{u}_i(k) \end{pmatrix}^T \underbrace{\begin{pmatrix} \boldsymbol{Q}_{11ij(k)} & \boldsymbol{Q}_{12ij(k)} & \boldsymbol{Q}_{13ij(k)} \\ * & \boldsymbol{Q}_{22ij(k)} & \boldsymbol{Q}_{23ij(k)} \\ * & * & \boldsymbol{Q}_{33ij(k)} \end{pmatrix}}_{\boldsymbol{Q}_{ij(k)}} \begin{pmatrix} \boldsymbol{x}_{ci}(k) \\ \hat{\boldsymbol{u}}_i(k) \\ \boldsymbol{u}_i(k) \end{pmatrix} \tag{2.26}$$

where

$$\boldsymbol{Q}_{11ij(k)} = \int_0^{h_{j(k)}} \boldsymbol{\Phi}_i^T(t)\boldsymbol{Q}_{ci}\boldsymbol{\Phi}_i(t)dt \tag{2.27a}$$

$$\boldsymbol{Q}_{12ij(k)} = \int_0^{\tau_{ij(k)}} \boldsymbol{\Phi}_i^T(t)\boldsymbol{Q}_{ci}\boldsymbol{\Gamma}_i(0,t)dt + \int_{\tau_{ij(k)}}^{h_{j(k)}} \boldsymbol{\Phi}_i^T(t)\boldsymbol{Q}_{ci}\boldsymbol{\Gamma}_i(0,\tau_{ij(k)})dt \tag{2.27b}$$

$$\boldsymbol{Q}_{13ij(k)} = \int_{\tau_{ij(k)}}^{h_{j(k)}} \boldsymbol{\Phi}_i^T(t)\boldsymbol{Q}_{ci}\boldsymbol{\Gamma}_i(\tau_{ij(k)},t)dt \tag{2.27c}$$

$$\boldsymbol{Q}_{22ij(k)} = \int_0^{\tau_{ij(k)}} \boldsymbol{\Gamma}_i^T(0,t)\boldsymbol{Q}_{ci}\boldsymbol{\Gamma}_i(0,t) + \boldsymbol{R}_{ci}dt + \int_{\tau_{ij(k)}}^{h_{j(k)}} \boldsymbol{\Gamma}_i^T(0,\tau_{ij(k)})\boldsymbol{Q}_{ci}\boldsymbol{\Gamma}_i(0,\tau_{ij(k)})dt \tag{2.27d}$$

$$\boldsymbol{Q}_{23ij(k)} = \int_{\tau_{ij(k)}}^{h_{j(k)}} \boldsymbol{\Gamma}_i^T(0,\tau_{ij(k)})\boldsymbol{Q}_{ci}\boldsymbol{\Gamma}_i(\tau_{ij(k)},t)dt \tag{2.27e}$$

$$\boldsymbol{Q}_{33ij(k)} = \int_{\tau_{ij(k)}}^{h_{j(k)}} \boldsymbol{\Gamma}_i^T(\tau_{ij(k)},t)\boldsymbol{Q}_{ci}\boldsymbol{\Gamma}_i(\tau_{ij(k)},t) + \boldsymbol{R}_{ci}dt \tag{2.27f}$$

with the matrices $\boldsymbol{\Phi}_i(t)$ defined according to (2.7a) and $\boldsymbol{\Gamma}_i(t_1,t_2) = \int_{t_1}^{t_2} e^{\boldsymbol{A}_{ci}(t-s)}ds\boldsymbol{B}_{ci}$.

Remark 2.8. The discretization of the continuous-time cost function (2.20) is first addressed in [San04, Paper B, Section 5.3]. It contains, however, some minor mistakes in its derivation which are corrected here.

Remark 2.9. Under the assumption that $\boldsymbol{Q}_{ci}$ and $\boldsymbol{R}_{ci}$ in (2.20) are symmetric and positive definite, it turns out that the weighting matrix $\boldsymbol{Q}_{ij(k)}$ is symmetric and positive semidefinite due to the equivalency between the discrete-time cost function (2.26) and the continuous-time cost function (2.20). The semidefiniteness of $\boldsymbol{Q}_{ij(k)}$ is because of its principal submatrix $\boldsymbol{Q}_{33ij(k)} = \boldsymbol{0}$ for $\delta_{ij(k)} = 0$.

Note that the resulting weighting matrix $\boldsymbol{Q}_{ij(k)}$ in (2.26) is *uncertain* time-varying due to its dependency on the uncertain time-varying input delay $\tau_i(k)$ for $\delta_{ij(k)} = 1$. Instead of considering $\tau_i(k)$ in the resulting cost function (2.26), we might consider the nominal input delay τ_{0i} which can be determined by taking the (weighted) average value of the network-induced transmission times. Replacing $\tau_i(k)$ in (2.27) by τ_{0i} yields a nominal cost function

$$J_{0i} = \sum_{k=0}^{\infty} \begin{pmatrix} \boldsymbol{x}_i(k) \\ \boldsymbol{u}_i(k) \end{pmatrix}^T \boldsymbol{Q}_{0ij(k)} \begin{pmatrix} \boldsymbol{x}_i(k) \\ \boldsymbol{u}_i(k) \end{pmatrix}. \tag{2.28}$$

The cost function assigned to the NCS is defined as the sum of the individual cost functions, i.e.

$$J = \sum_{i=1}^{M} J_i = \sum_{k=0}^{\infty} \begin{pmatrix} \boldsymbol{x}(k) \\ \boldsymbol{u}(k) \end{pmatrix}^T \boldsymbol{Q}_{j(k)} \begin{pmatrix} \boldsymbol{x}(k) \\ \boldsymbol{u}(k) \end{pmatrix} \tag{2.29}$$

or in case of considering nominal input delays τ_{0i} as

$$J_0 = \sum_{i=1}^{M} J_{0i} = \sum_{k=0}^{\infty} \begin{pmatrix} \boldsymbol{x}(k) \\ \boldsymbol{u}(k) \end{pmatrix}^T \boldsymbol{Q}_{0j(k)} \begin{pmatrix} \boldsymbol{x}(k) \\ \boldsymbol{u}(k) \end{pmatrix} \tag{2.30}$$

where the state vector $\boldsymbol{x}(k)$ and the control vector $\boldsymbol{u}(k)$ are defined according to (2.11) and the weighting matrices $\boldsymbol{Q}_{j(k)}$ or $\boldsymbol{Q}_{0j(k)}$ are resulted by construction.

Remark 2.10. The nominal cost function (2.30) quantifies the variation of the nominal input delay $\tau_{0j(k)}$ and the variation of the sampling interval $h_{j(k)}$ with respect to $j(k) \in \mathbb{M}$. It does not, however, quantify the variation of the actual input delay $\tau_i(k)$ around its nominal value. Regarding the actual input delay in the cost function will introduce unnecessary complexity in the codesign procedure later on for a possibly marginal benefit.

Remark 2.11. Alternative to the nominal cost function (2.30), an overapproximation of the cost function (2.29) using the Taylor series expansion method can also be considered as already shown in [AGL11].

In the following, the nominal cost function (2.30) is considered as an objective function in the codesign process. The cost function (2.29) is only considered for evaluation purposes and not in the codesign process.

3 Codesign Problem

The main control and scheduling codesign problem is introduced in this chapter. To this end, the structure of the control law is first defined. Based on the NCS model derived in Chapter 2 and the defined control law, the main control and scheduling codesign problem is formulated as an optimization problem with the objective of minimizing the cost function introduced in Chapter 2. The properties of the resulting optimization problem in terms of convexity, tractability, and stability implication are discussed.

3.1 Problem Formulation

For each control task T_i, we consider a state feedback control law

$$\boldsymbol{u}_i(k) = \boldsymbol{K}_i(k)\boldsymbol{x}_i(k) \tag{3.1}$$

with the time-varying feedback matrix $\boldsymbol{K}_i(k) \in \mathbb{R}^{m_i \times (n_i+m_i)}$. Substituting the control law (3.1) into the NCS model (2.18) results in a closed-loop switched polytopic system

$$\boldsymbol{x}(k+1) = \underbrace{\left(\sum_{\ell=1}^{L+1} \mu_\ell(\rho_{j(k)}) \tilde{\boldsymbol{A}}_{j(k)\ell}(k) + \boldsymbol{D}_{j(k)} \boldsymbol{E}_{j(k)} \tilde{\boldsymbol{F}}(k) \right)}_{\mathcal{A}_{j(k)}(k)} \boldsymbol{x}(k) \tag{3.2}$$

where

$$\tilde{\boldsymbol{A}}_{j(k)\ell}(k) = \boldsymbol{A}_{j(k)\ell} + \boldsymbol{B}_{j(k)\ell}\boldsymbol{K}(k) \tag{3.3a}$$

$$\tilde{\boldsymbol{F}}(k) = \boldsymbol{F}_{\mathrm{a}} + \boldsymbol{F}_{\mathrm{b}}\boldsymbol{K}(k) \tag{3.3b}$$

$$\boldsymbol{K}(k) = \operatorname{diag}\big(\boldsymbol{K}_1(k), \ldots, \boldsymbol{K}_M(k)\big). \tag{3.3c}$$

Substituting further (3.1) into the cost function (2.30) yields a closed-loop cost function

$$J_0 = \sum_{k=0}^{\infty} \boldsymbol{x}^T(k) \tilde{\boldsymbol{Q}}_{0j(k)} \boldsymbol{x}(k) \tag{3.4}$$

where

$$\tilde{\boldsymbol{Q}}_{0j(k)} = \begin{pmatrix} \boldsymbol{I} \\ \boldsymbol{K}(k) \end{pmatrix}^T \boldsymbol{Q}_{0j(k)} \begin{pmatrix} \boldsymbol{I} \\ \boldsymbol{K}(k) \end{pmatrix}. \tag{3.5}$$

The main control and scheduling codesign problem can now be formulated as

Problem 3.1 *For the closed-loop switched polytopic system* (3.2) *find the optimal feedback sequence* $\boldsymbol{K}^*(0),\ldots,\boldsymbol{K}^*(\infty)$ *and switching sequence* $j^*(0),\ldots,j^*(\infty)$ *such that the closed-loop cost function* (3.4) *is robustly minimized for all* $\tau_{j(k)} \in [\underline{\tau}_{j(k)}, \overline{\tau}_{j(k)}]$, $k \in \mathbb{N}_0$, *i.e.*

$$\min_{\substack{j(0),\ldots,j(\infty)\\ \boldsymbol{K}(0),\ldots,\boldsymbol{K}(\infty)}} J_0^{\max} \quad \text{subject to (3.2)} \tag{3.6}$$

where

$$J_0^{\max} \triangleq \max_{\tau_{j(0)},\ldots,\tau_{j(\infty)}} J_0. \tag{3.7}$$

3.2 Problem Properties

Problem 3.1 is a robust optimization problem (minimax problem). The term robust optimization refers to an optimization problem with uncertainty in the problem data. The uncertainty here is in the prediction of system dynamics based on (3.2). For a given initial value $\boldsymbol{x}(0)$, the predicted states $\boldsymbol{x}(k)$, $\forall k \in \mathbb{N}$, define polytopic sets instead of exactly known values.

Lemma 3.1 *The objective function* (3.7) *for the switched polytopic system* (3.2) *and a given switching sequence* $j(0),\ldots,j(\infty)$ *is convex in the uncertain input delay sequence* $\tau_{j(0)},\ldots,\tau_{j(\infty)}$.

PROOF. The quadratic cost function J_0 in (3.4) is a convex function in $\boldsymbol{x}(k)$. Furthermore, the uncertain input delay $\tau_{j(k)}$, $\forall k \in \mathbb{N}_0$, generates for a given switching sequence a convex set of state predictions based on (3.2). In [Löf03, Theorem 4.1], it has been shown that for a convex polytopic set $\mathbb{X}_{\text{conv}}$ and a convex function $f : \mathbb{X}_{\text{conv}} \to \mathbb{R}$ the maximum value of f is attained at one of the vertices of the polytope $\mathbb{X}_{\text{conv}}$. This implies that the objective function (3.7) is convex in $\tau_{j(0)},\ldots,\tau_{j(\infty)}$, completing the proof. □

Even though Problem 3.1 is convex, it is unfortunately computationally intractable. The intractability can be seen from its scheduling and/or uncertain input delay dependency. From the scheduling point of view, the number of possible switching sequences grows exponentially in the time-horizon, making Problem 3.1 intractable even for a finite time-horizon cost function. Within the time interval $k = 0$ to $k = N$, for instance, the number of possible switching sequences $j(0),\ldots,j(N)$ to be inspected for optimization is equal to M^N with M denoting the number of subsystems. From the uncertain input delay perspective, one can notice that the number of vertices of the polytopic set that defines the predicted state $\boldsymbol{x}(k)$ grows exponentially with k as well. It is worth to note that this exponential behavior is one of the main obstacles in minimax optimization problems.

The solvability of Problem 3.1 implies the global uniform asymptotic stability (GUAS) of the closed-loop discrete-time switched polytopic system (3.2). This is due to the fact that the objective function (3.7) has a finite bound if and only if the switched polytopic system (3.2) is asymptotically stable for all initial state values $\boldsymbol{x}(0) \in \mathbb{R}^{n+m}$. However, since the dynamics of each plant P_i in the NCS evolves according to (2.1) in a continuous-time manner, the inter-sample behavior is also of particular importance. In the next theorem, we prove that the GUAS of (3.2) implies the GUAS of the NCS taking the continuous-time dynamics (2.1) into account. This means that the inter-sample behavior is bounded and also converges toward the origin.

Lemma 3.2 *Assume the GUAS of the discrete-time switched polytopic system* (3.2) *for a given switching sequence* $j(0), \ldots, j(\infty)$ *and feedback sequence* $\boldsymbol{K}(0), \ldots, \boldsymbol{K}(\infty)$. *The NCS shown in Figure 2.1 with the plant dynamics given by* (2.1) *is then globally uniformally asymptotically stable.*

Proof. It has been shown in Section 2.4 that the cost function (2.26) is equivalent to the continuous-time quadratic cost function (2.20). This means that the cost function (2.26) accounts also for the inter-sample behavior. To prove this lemma, therefore, we just need to show that the cost function (2.29) has a finite bound. To this end, the cost function (2.29) is reformulated, cf. [AGL12b], as

$$\begin{aligned} J &= \sum_{k=0}^{\infty} \begin{pmatrix} \boldsymbol{x}(k) \\ \boldsymbol{u}(k) \end{pmatrix}^T \left(\boldsymbol{Q}_{0j(k)} + \Delta \boldsymbol{Q}_{j(k)} \right) \begin{pmatrix} \boldsymbol{x}(k) \\ \boldsymbol{u}(k) \end{pmatrix} \\ &= J_0 + \Delta J. \end{aligned} \tag{3.8}$$

GUAS of the switched polytopic system (3.2) implies that the first part J_0 has a finite bound. For the second part ΔJ, its stage cost approaches zero as $k \to \infty$ since

$$\begin{aligned} \lim_{k \to \infty} \boldsymbol{x}(k) &= \boldsymbol{0} \\ \lim_{k \to \infty} \boldsymbol{u}(k) &= \boldsymbol{0}. \end{aligned} \tag{3.9}$$

This means that the second part ΔJ is also bounded, implying that the cost function (2.29) is bounded. This completes the proof. □

For obtaining a tractable optimization problem from Problem 3.1, five suboptimal approaches are proposed during the doctoral studies. A detailed description of each of them as well as a comparison among them from performance and complexity perspectives is given in the following chapters.

4 Periodic Control and Scheduling (PCS)

In this chapter, a periodic control and scheduling codesign strategy is proposed as a suboptimal solution to Problem 3.1. According to the discussion in Section 3.2, the intractability of Problem 3.1 seen from the scheduling aspect is resolved by imposing periodicity on the switching sequence $j(0), \ldots, j(\infty)$ and thus on the feedback sequence $\boldsymbol{K}(0), \ldots, \boldsymbol{K}(\infty)$. By doing this, a minimax optimization problem where a finite set rather than an infinite set of switching and control variables is to be determined. The imposed periodicity might, however, impair the resulting control performance. In order to improve the performance, the resulting minimax optimization problem over an infinite time-horizon is solved at each time instant t_k for the current state $\boldsymbol{x}(k)$. Hence, the imposed periodicity is suppressed while improving the control performance.

4.1 Problem Formulation

Before delving into the details of the periodic control and scheduling codesign problem, the following definitions are in order.

Definition 4.1 *A p-periodic switching-feedback sequence σ is defined by*

$$\sigma \triangleq \big(\underbrace{\big(j(0), \boldsymbol{K}(0)\big)}_{\sigma(0)}, \ldots, \underbrace{\big(j(p-1), \boldsymbol{K}(p-1)\big)}_{\sigma(p-1)}\big) \tag{4.1}$$

which implies for all $k \in \mathbb{N}_0$ that

$$j(k+p) = j(k) \tag{4.2a}$$
$$\boldsymbol{K}(k+p) = \boldsymbol{K}(k). \tag{4.2b}$$

Definition 4.2 *A p-periodic switching-feedback sequence σ chosen at time instant t_k is denoted with abuse of notation as σ_k.*

Definition 4.3 *The set of admissible (i.e. stabilizing the switched polytopic system (3.2)) p-periodic switching-feedback sequences σ is defined by*

$$\mathbb{S}_p = \{\sigma \mid j(k+p) = j(k),\ \boldsymbol{K}(k+p) = \boldsymbol{K}(k), \text{ and (3.2) is GUAS}\}. \tag{4.3}$$

Consider a discrete-time quadratic cost function

$$J_p(k) = \sum_{i=0}^{\infty} \boldsymbol{x}^T(k+i)\tilde{\boldsymbol{Q}}_{0j(k+i)}\boldsymbol{x}(k+i) \tag{4.4}$$

with the weighting matrix $\tilde{\boldsymbol{Q}}_{0j(k+i)}$ defined according to (3.5) and the p-periodic switching indices $j(k+i)$ and feedback matrices $\boldsymbol{K}(k+i)$, $\forall i \in \mathbb{N}_0$, satisfying (4.2). The periodic control and scheduling (PCS) codesign problem can now be formulated as

Problem 4.1 *For the closed-loop switched polytopic system (3.2) and the current state $\boldsymbol{x}(k)$ find the optimal p-periodic switching-feedback sequence $\sigma_k^* \in \mathbb{S}_p$ such that the closed-loop cost function (4.4) is robustly minimized for all $\tau_{j(k)} \in [\underline{\tau}_{j(k)}, \overline{\tau}_{j(k)}]$, i.e.*

$$\min_{\sigma \in \mathbb{S}_p} J_p^{\max}(k) \qquad \text{subject to (3.2)} \tag{4.5}$$

where

$$J_p^{\max}(k) \triangleq \max_{\tau_{j(k)},\dots,\tau_{j(\infty)}} J_p(k). \tag{4.6}$$

Apply then the first element of σ_k^*, i.e. $\sigma_k(0) = \big(j^*(k), \boldsymbol{K}^*(k)\big)$, to the closed-loop switched polytopic system (3.2) and reoptimize at next time instant t_{k+1}.

Remark 4.1. Due to the imposed periodicity on the switching sequence $j(k), \dots, j(\infty)$ and the feedback sequence $\boldsymbol{K}(k), \dots, \boldsymbol{K}(\infty)$, a finite set of switching and feedback variables is to be determined. Problem 4.1 is still, however, computationally intractable due to the dependency of its objective function (4.6) on the uncertain time-varying input delay $\tau_{j(k)}$. This issue has been addressed in the literature yet for linear systems, see e.g. [KBM96, Section 3.1], and solved by deriving an upper bound on (4.6) and considering it as a new objective function to get a tractable optimization problem.

4.2 Upper Bound Derivation

Consider a p-periodic Lyapunov function

$$V(k) = \boldsymbol{x}^T(k)\boldsymbol{P}_\sigma(\kappa)\boldsymbol{x}(k), \quad \kappa = (k - k_0) \bmod p \tag{4.7}$$

where the p-periodic Lyapunov matrices $\boldsymbol{P}_\sigma(\kappa) \in \mathbb{R}^{(n+m)\times(n+m)}$ are symmetric and positive definite. The subindex σ indicates the affiliation to a specific p-periodic switching-feedback sequence $\sigma \in \mathbb{S}_p$ while k_0 the index of initial time instant.

Remark 4.2. Note that the chosen Lyapunov function (4.7) is positive definite, decrescent and radially unbounded since

$$\alpha_1 \left\| \boldsymbol{x}(k) \right\|_2^2 \le \boldsymbol{x}^T(k) \boldsymbol{P}_\sigma(\kappa) \boldsymbol{x}(k) \le \alpha_2 \left\| \boldsymbol{x}(k) \right\|_2^2 \tag{4.8}$$

for all $\boldsymbol{x}(k) \in \mathbb{R}^{n+m}$ and $k \in \mathbb{N}_0$ with

$$\alpha_1 := \min_{\kappa \in \{0,1,\dots,p-1\}} \lambda_{\min}\big(\boldsymbol{P}_\sigma(\kappa)\big) > 0, \quad \alpha_2 := \max_{\kappa \in \{0,1,\dots,p-1\}} \lambda_{\max}\big(\boldsymbol{P}_\sigma(\kappa)\big) > 0. \tag{4.9}$$

The difference of the Lyapunov function, i.e. $\Delta V(k) \triangleq V(k+1) - V(k)$, along trajectories of the closed-loop switched polytopic system (3.2) is given by

$$\Delta V(k) = \boldsymbol{x}^T(k) \big(\boldsymbol{\mathcal{A}}_{j(\kappa)}^T(k) \boldsymbol{P}_\sigma(\kappa^+) \boldsymbol{\mathcal{A}}_{j(\kappa)}(k) - \boldsymbol{P}_\sigma(\kappa) \big) \boldsymbol{x}(k) \tag{4.10}$$

where

$$\kappa^+ = (\kappa + 1) \bmod p. \tag{4.11}$$

Theorem 4.1 *For a p-periodic switching-feedback sequence $\sigma \in \mathbb{S}_p$ given at time instant t_k, the objective function (4.6) is upper bounded by*

$$J_p^{\max}(k) < \boldsymbol{x}^T(k) \boldsymbol{P}_\sigma(0) \boldsymbol{x}(k) \tag{4.12}$$

provided that the difference of the Lyapunov function (4.10) satisfies

$$\Delta V(k) < -\boldsymbol{x}^T(k) \tilde{\boldsymbol{Q}}_{0j(k)} \boldsymbol{x}(k) \tag{4.13}$$

for all $\boldsymbol{x}(k) \in \mathbb{R}^{n+m} \backslash \{\boldsymbol{0}\}$ with the weighting matrix $\tilde{\boldsymbol{Q}}_{0j(k)}$ defined according to (3.5).

Proof. Summing the inequality (4.13) over $k, \dots, \infty$ yields

$$\lim_{k \to \infty} V(k) - V(k) < -\sum_{i=0}^{\infty} \boldsymbol{x}^T(k+i) \tilde{\boldsymbol{Q}}_{0j(k+i)} \boldsymbol{x}(k+i). \tag{4.14}$$

The satisfaction of (4.13) implies the negative definiteness of $\Delta V(k)$ in (4.10). Thus the Lyapunov function $V(k)$ approaches the origin as $k \to \infty$ due to its positive definiteness assumption. Equation (4.14) implies thus that

$$J_p^{\max}(k) < \boldsymbol{x}^T(k) \boldsymbol{P}_\sigma(\kappa) \boldsymbol{x}(k). \tag{4.15}$$

The first element $\sigma(0)$ of the given p-periodic switching-feedback sequence σ is applied at time instant t_k, i.e. $k = k_0$. This corresponds to the pointer value $\kappa = 0$ and hence

$$\begin{aligned} J_p^{\max}(k) &< \boldsymbol{x}^T(k) \boldsymbol{P}_\sigma(0) \boldsymbol{x}(k) \\ &\le \lambda_{\max}\big(\boldsymbol{P}_\sigma(0)\big) \boldsymbol{x}^T(k) \boldsymbol{x}(k) \\ &< \operatorname{tr}\big(\boldsymbol{P}_\sigma(0)\big) \boldsymbol{x}^T(k) \boldsymbol{x}(k) \end{aligned} \tag{4.16}$$

where the second non-strict inequality follows from the Rayleigh-Ritz inequality [Mey00, Example 7.5.1] and the last inequality due to the fact that the $\mathrm{tr}(\cdot)$ corresponds to the sum of the eigenvalues and $\boldsymbol{P}_\sigma(0) \succ \boldsymbol{0}$, $\forall \sigma \in \mathbb{S}_p$. This completes the proof. □

Problem 4.1 can now be modified based on Theorem 4.1 as a computationally tractable periodic control and scheduling codesign problem

Problem 4.2 *For the closed-loop switched polytopic system* (3.2) *and the current state* $\boldsymbol{x}(k)$ *find the optimal p-periodic switching-feedback sequence* $\sigma_k^* \in \mathbb{S}_p$ *such that the upper bound* (4.12) *is minimized, i.e.*

$$\min_{\sigma \in \mathbb{S}_p} \boldsymbol{x}^T(k)\boldsymbol{P}_\sigma(0)\boldsymbol{x}(k) \quad \text{subject to (4.13).} \tag{4.17}$$

Apply then the first element of σ_k^*, *i.e.* $\sigma_k(0) = \big(j^*(k), \boldsymbol{K}^*(k)\big)$, *to the closed-loop switched polytopic system* (3.2) *and reoptimize at next time instant* t_{k+1}.

Remark 4.3. The consideration of an upper bound on the "worst-case" cost function (4.6) is a classical approach used in the model predictive control community, see e.g. [XS93, KBM96] for linear systems and [BRRA09] for periodically switched systems. The resulting solution of Problem 4.2 can generally be regarded due to the upper bound consideration as a sub-optimal solution to Problem 4.1. An analytical assessment of the amount of sub-optimality is still open even in the model predictive control community.

Problem 4.2 is solved in the following in two steps. In the first step, the set $\mathbb{S}_p$ of p-periodic switching-feedback sequences σ that satisfies (4.13) is determined. Since the switching index $j(k)$ belongs to the finite set $\mathbb{M}$, all p-periodic switching sequences satisfying (4.2a) can be determined by permutations with repetition. The corresponding p-periodic feedback matrices $\boldsymbol{K}(k)$ satisfying (4.2b) and the stability condition (4.13) are determined from an LMI optimization problem as discussed later on. Once we determine the set $\mathbb{S}_p$, we run the second step by getting through all of the p-periodic switching-feedback sequence candidates $\sigma \in \mathbb{S}_p$, i.e. exhaustive search, and checking whether each candidate satisfies the problem's statement.

Remark 4.4. The two-step procedure of solving Problem 4.2 is adopted from a procedure proposed in [Gör12, Chapter 4] for periodic control and scheduling of switched linear systems (1.5). This procedure is generalized here for periodic control and scheduling of the switched polytopic system with additive norm-bounded uncertainty (3.2).

4.3 Periodic Control

Theorem 4.2 *For a given p-periodic switching sequence* $j(0), \ldots, j(p-1)$ *satisfying* (4.2a), *the corresponding p-periodic feedback sequence* $\boldsymbol{K}(0), \ldots, \boldsymbol{K}(p-1)$ *satisfying*

(4.2b) *and the stability condition* (4.13) *results from the LMI optimization problem*

$$\min_{G(\kappa),\, W(\kappa),\, Z_\sigma(\kappa),\, \beta_\ell(\kappa)} \operatorname{tr}\big(P_\sigma(0)\big) \quad \text{subject to} \tag{4.18a}$$

$$\begin{pmatrix} G^T(\kappa) + G(\kappa) - Z_\sigma(\kappa) & * & * & * \\ F_a G(\kappa) + F_b W(\kappa) & \beta_\ell(\kappa) I & * & * \\ \sqrt{Q_{0j(\kappa)}} \begin{pmatrix} G(\kappa) \\ W(\kappa) \end{pmatrix} & 0 & I & * \\ A_{j(\kappa)\ell} G(\kappa) + B_{j(\kappa)\ell} W(\kappa) & 0 & 0 & Z_\sigma(\kappa^+) - \beta_\ell(\kappa) D_{j(\kappa)} D_{j(\kappa)}^T \end{pmatrix} \succ 0 \tag{4.18b}$$

for all $\ell \in \{1, \ldots, L+1\}$ *and* $\kappa \in \{0, 1, \ldots, p-1\}$ *with the LMI variables: Block-diagonal matrices* $G(\kappa) \in \mathbb{R}^{(n+m)\times(n+m)}$ *and* $W(\kappa) \in \mathbb{R}^{m\times(n+m)}$, *symmetric and positive definite matrices* $Z_\sigma(\kappa) \in \mathbb{R}^{(n+m)\times(n+m)}$, *and positive real scalars* $\beta_\ell(\kappa) \in \mathbb{R}^+$. *The corresponding p-periodic feedback matrices in* (4.2b) *and the p-periodic Lyapunov matrices in* (4.7) *are given by*

$$K(\kappa) = W(\kappa) G^{-1}(\kappa) \tag{4.19a}$$

$$P_\sigma(\kappa) = Z_\sigma^{-1}(\kappa). \tag{4.19b}$$

PROOF. Let's assume that the LMI constraints (4.18b) are feasible. Then, from the first principal submatrix we obtain

$$G^T(\kappa) + G(\kappa) - Z_\sigma(\kappa) \succ 0 \tag{4.20}$$

or equivalently

$$G^T(\kappa) + G(\kappa) \succ Z_\sigma(\kappa) \succ 0. \tag{4.21}$$

Thus, $G(\kappa)$ is of full rank for which its inverse always exists. Furthermore, it holds that

$$Z_\sigma^{-1}(\kappa) \succ 0 \tag{4.22}$$

since an inversion does not affect definiteness. Pre- and post-multiplying (4.22) by $(Z_\sigma(\kappa) - G(\kappa))^T$ and its transpose, respectively, yields [DB01]

$$G^T(\kappa) Z_\sigma^{-1}(\kappa) G(\kappa) \succeq G^T(\kappa) + G(\kappa) - Z_\sigma(\kappa). \tag{4.23}$$

Therefore, (4.18b) implies

$$\begin{pmatrix} G^T(\kappa) Z_\sigma^{-1}(\kappa) G(\kappa) & * & * & * \\ F_a G(\kappa) + F_b W(\kappa) & \beta_\ell(\kappa) I & * & * \\ \sqrt{Q_{0j(\kappa)}} \begin{pmatrix} G(\kappa) \\ W(\kappa) \end{pmatrix} & 0 & I & * \\ A_{j(\kappa)\ell} G(\kappa) + B_{j(\kappa)\ell} W(\kappa) & 0 & 0 & Z_\sigma(\kappa^+) - \beta_\ell(\kappa) D_{j(\kappa)} D_{j(\kappa)}^T \end{pmatrix} \succ 0 \tag{4.24}$$

Substituting $\boldsymbol{W}(\kappa) = \boldsymbol{K}(\kappa)\boldsymbol{G}(\kappa)$ into (4.24) and further pre- and post-multiplying it by $\mathrm{diag}(\boldsymbol{G}^{-T}(\kappa), \boldsymbol{I}, \boldsymbol{I}, \boldsymbol{I})$ and its transpose, respectively, results in

$$\begin{pmatrix} \boldsymbol{P}_\sigma(\kappa) & * & * & * \\ \tilde{\boldsymbol{F}}(\kappa) & \beta_\ell(\kappa)\boldsymbol{I} & * & * \\ \sqrt{\boldsymbol{Q}_{0j(\kappa)}}\begin{pmatrix} \boldsymbol{I} \\ \boldsymbol{K}(\kappa) \end{pmatrix} & 0 & \boldsymbol{I} & * \\ \tilde{\boldsymbol{A}}_{j(\kappa)\ell}(\kappa) & 0 & 0 & \boldsymbol{Z}_\sigma(\kappa^+) - \beta_\ell(\kappa)\boldsymbol{D}_{j(\kappa)}\boldsymbol{D}^T_{j(\kappa)} \end{pmatrix} \succ \boldsymbol{0}. \tag{4.25}$$

with $\tilde{\boldsymbol{F}}(\kappa)$ and $\tilde{\boldsymbol{A}}_{j(\kappa)\ell}(\kappa)$ defined according to (3.3). Applying the Schur complement twice to (4.25) results in

$$\begin{pmatrix} \boldsymbol{P}_\sigma(\kappa) - \tilde{\boldsymbol{Q}}_{0j(\kappa)} - \beta_\ell^{-1}(\kappa)\tilde{\boldsymbol{F}}^T(\kappa)\tilde{\boldsymbol{F}}(\kappa) & * \\ \tilde{\boldsymbol{A}}_{j(\kappa)\ell}(\kappa) & \boldsymbol{Z}_\sigma(\kappa^+) - \beta_\ell(\kappa)\boldsymbol{D}_{j(\kappa)}\boldsymbol{D}^T_{j(\kappa)} \end{pmatrix} \succ \boldsymbol{0} \tag{4.26}$$

with $\tilde{\boldsymbol{Q}}_{0j(\kappa)}$ defined according to (3.5). Equation (4.26) can be rewritten as

$$\boldsymbol{Y}_\ell(\kappa) - \beta_\ell(\kappa)\boldsymbol{M}_{j(\kappa)}\boldsymbol{M}^T_{j(\kappa)} - \beta_\ell^{-1}(\kappa)\boldsymbol{N}^T(\kappa)\boldsymbol{N}(\kappa) \succ \boldsymbol{0} \tag{4.27}$$

with

$$\boldsymbol{Y}_\ell(\kappa) = \begin{pmatrix} \boldsymbol{P}_\sigma(\kappa) - \tilde{\boldsymbol{Q}}_{0j(\kappa)} & * \\ \tilde{\boldsymbol{A}}_{j(\kappa)\ell}(\kappa) & \boldsymbol{Z}_\sigma(\kappa^+) \end{pmatrix}, \quad \boldsymbol{M}_{j(\kappa)} = \begin{pmatrix} \boldsymbol{0} \\ \boldsymbol{D}_{j(\kappa)} \end{pmatrix}, \quad \boldsymbol{N}(\kappa) = \begin{pmatrix} \tilde{\boldsymbol{F}}(\kappa) & \boldsymbol{0} \end{pmatrix}.$$

Based on Lemma A.2, inequality (4.27) is equivalent to

$$\begin{pmatrix} \boldsymbol{P}_\sigma(\kappa) - \tilde{\boldsymbol{Q}}_{0j(\kappa)} & * \\ \tilde{\boldsymbol{A}}_{j(\kappa)\ell}(\kappa) + \boldsymbol{D}_{j(\kappa)}\boldsymbol{E}_{j(\kappa)}\tilde{\boldsymbol{F}}(\kappa) & \boldsymbol{Z}_\sigma(\kappa^+) \end{pmatrix} \succ \boldsymbol{0}. \tag{4.28}$$

Multiplying (4.28) by $\mu_\ell(\rho_{j(\kappa)})$ and further summing it over $\ell = 1, \ldots, L+1$ yields

$$\begin{pmatrix} \boldsymbol{P}_\sigma(\kappa) - \tilde{\boldsymbol{Q}}_{0j(\kappa)} & * \\ \boldsymbol{\mathcal{A}}_{j(\kappa)}(k) & \boldsymbol{Z}_\sigma(\kappa^+) \end{pmatrix} \succ \boldsymbol{0}. \tag{4.29}$$

Applying the Schur complement to (4.29) results in

$$\boldsymbol{P}_\sigma(\kappa) - \tilde{\boldsymbol{Q}}_{0j(\kappa)} - \boldsymbol{\mathcal{A}}^T_{j(\kappa)}(k)\boldsymbol{P}_\sigma(\kappa^+)\boldsymbol{\mathcal{A}}_{j(\kappa)}(k) \succ \boldsymbol{0}. \tag{4.30}$$

which corresponds to the stability condition (4.13). For the objective function of (4.18a), it is introduced for the purpose of minimizing the upper bound (4.16). Other objective functions, e.g. $\lambda_{\max}\big(\boldsymbol{P}_\sigma(0)\big)$, can be considered as well. This completes the proof. □

Remark 4.5. The LMI optimization problem (4.18) can be solved completely offline for each admissible p-periodic switching sequence satisfying (4.2a). The involved offline computational complexity of (4.18) is generally characterized by $|\mathbb{S}_p| \cong M^p$ and hence it grows exponentially with the period p.

Remark 4.6. The LMI optimization problem (4.18) can be solved using CVX, a package for specifying and solving convex programs [GB14, GB08]. Alternatively, we can use the MATLAB toolbox YALMIP [49] with the SeDuMi solver [50]. Since the $\mathrm{tr}(\cdot)$ operator is not explicitly defined in YALMIP, however, the $\mathrm{logdet}(\cdot)$ operator can be considered instead. This consideration is motivated by the fact that $\mathrm{tr}(\cdot)$ corresponds to the sum of the eigenvalues and $\mathrm{logdet}(\cdot)$ corresponds to the sum of the logarithmized eigenvalues.

Remark 4.7. According to Theorem 4.2, a one-step cyclical shift of a given p-periodic switching sequence $j(0), j(1), \ldots, j(p-1)$, i.e.

$$j(1), j(2), \ldots, j(p-1), j(0) \tag{4.31}$$

leads to the same p-periodic feedback matrices $\boldsymbol{K}(0), \ldots, \boldsymbol{K}(p-1)$ and Lyapunov matrices $\boldsymbol{P}_\sigma(0), \ldots, \boldsymbol{P}_\sigma(p-1)$ but shifted cyclically by one-step. From the Lyapunov matrices perspective, this means that

$$\boldsymbol{P}_{\overline{\sigma}}(\kappa) = \boldsymbol{P}_\sigma(\kappa^+) \tag{4.32}$$

with the one-step cyclically shifted p-periodic switching-feedback sequence $\overline{\sigma}$ given by

$$\overline{\sigma} \triangleq \big(\sigma(1), \ldots, \sigma(p-1), \sigma(0)\big) \tag{4.33}$$

and κ^+ defined according to (4.11). Equation (4.32) holds due to the fact that the LMI constraints (4.18b) for a p-periodic switching sequence and its one-step cyclically shifted version are identical. This property will be used during the discussion of the possible solutions to Problem 4.2 in the following sections.

Remark 4.8. Based on Remark 4.7, one can conclude that if a given p-periodic switching-feedback sequence $\sigma \in \mathbb{S}_p$ stabilizes the switched polytopic system (3.2), i.e. admissible, then all its cyclic versions $\overline{\sigma}$ are also admissible and thus they must belong to the set $\mathbb{S}_p$.

4.4 Solution based on Exhaustive Search

A trivial solution to Problem 4.2, with the elements of the set $\mathbb{S}_p$ resulting from Theorem 4.2, consists in systematically enumerating all of the p-periodic switching-feedback sequence candidates $\sigma \in \mathbb{S}_p$ and checking whether each candidate satisfies the problem's statement. This can be formalized as

Theorem 4.3 *Problem 4.2 with the elements of the set $\mathbb{S}_p$ resulting from Theorem 4.2 is solved for the current state $\boldsymbol{x}(k)$ based on exhaustive search, i.e.*

$$\sigma_k^* = \arg\min_{\sigma \in \mathbb{S}_p} \boldsymbol{x}^T(k) \boldsymbol{P}_\sigma(0) \boldsymbol{x}(k) \tag{4.34}$$

which corresponds to

$$\boldsymbol{x}^T(k) \boldsymbol{P}_{\sigma_k^*}(0) \boldsymbol{x}(k) \leq \boldsymbol{x}^T(k) \boldsymbol{P}_\sigma(0) \boldsymbol{x}(k) \quad \forall \sigma \in \mathbb{S}_p\,. \tag{4.35}$$

Moreover, the closed-loop switched polytopic system (3.2) under the periodic codesign strategy (4.34) is GUAS.

Proof. Based on (4.34), the optimal p-periodic switching-feedback sequence at initial time instant t_0 is given by

$$\sigma_0^* = \arg\min_{\sigma\in\mathbb{S}_p} \boldsymbol{x}^T(0)\boldsymbol{P}_\sigma(0)\boldsymbol{x}(0). \tag{4.36}$$

A reoptimization is then performed at next time instant t_1, leading to

$$\sigma_1^* = \arg\min_{\sigma\in\mathbb{S}_p} \boldsymbol{x}^T(1)\boldsymbol{P}_\sigma(0)\boldsymbol{x}(1). \tag{4.37}$$

The one-step cyclic shift of σ_0^*, i.e. $\overline{\sigma}_0^*$ defined in (4.33), is one of the possible candidates that can be chosen at time instant t_1. Regarding Remark 4.7, we obtain

$$\boldsymbol{x}^T(1)\boldsymbol{P}_{\sigma_1^*}(0)\boldsymbol{x}(1) \leq \boldsymbol{x}^T(1)\boldsymbol{P}_{\overline{\sigma}_0^*}(0)\boldsymbol{x}(1) < \boldsymbol{x}^T(0)\boldsymbol{P}_{\sigma_0^*}(0)\boldsymbol{x}(0). \tag{4.38}$$

where the first inequality is fulfilled due to the reoptimization and the second inequality due to the stability condition (4.13) while keeping (4.32) in mind. The same conclusion can be made at each time instant t_k, i.e.

$$\boldsymbol{x}^T(k)\boldsymbol{P}_{\sigma_k^*}(0)\boldsymbol{x}(k) < \boldsymbol{x}^T(k-1)\boldsymbol{P}_{\sigma_{k-1}^*}(0)\boldsymbol{x}(k-1) \quad \forall k \in \mathbb{N}. \tag{4.39}$$

Equation (4.39) implies that $\lim_{k\to\infty} \boldsymbol{x}(k) = \boldsymbol{0}$, proving the GUAS of the closed-loop switched polytopic system (3.2). □

Remark 4.9. The online computational complexity of the periodic codesign strategy (4.34) is characterized by $|\mathbb{S}_p| \cong M^p$. Such exponential complexity with respect to the period p might lead to a large amount of scheduling overhead, impairing instead of improving the control performance via dynamic scheduling. Reduction of the online complexity is thus crucial. Two relaxed versions of the periodic codesign strategy (4.34) are therefore proposed in the following with a significantly less online complexity at the cost of introducing further suboptimality in the resulting solutions.

Remark 4.10. The worst-case cost resulting for the periodic codesign strategy (4.34) is upper bounded by

$$J_p^{\max}(k) < \min_{\sigma\in\mathbb{S}_p} \boldsymbol{x}^T(k)\boldsymbol{P}_\sigma(0)\boldsymbol{x}(k) \quad \forall k \in \mathbb{N}_0. \tag{4.40}$$

From (4.39), furthermore, one can easily notice that the optimal upper bound resulting at time instant t_k is always less than the optimal one resulting at last time instant t_{k-1}.

4.5 Solution based on Relaxation

The main idea in this section is to extract a relaxed set $\hat{\mathbb{S}}_p \subset \mathbb{S}_p$, reducing thus the online complexity while preserving closed-loop stability. Problem 4.2 is then solved based on Theorem 4.3 with the set $\mathbb{S}_p$ in (4.34) replaced by the relaxed set $\hat{\mathbb{S}}_p$.

Theorem 4.4 *The solution to Problem 4.2 for the current state $\boldsymbol{x}(k)$ and the relaxed set $\hat{\mathbb{S}}_p \subset \mathbb{S}_p$ is given by*

$$\sigma_k^* = \arg\min_{\hat{\sigma}\in\hat{\mathbb{S}}_p} \boldsymbol{x}^T(k)\boldsymbol{P}_{\hat{\sigma}}(0)\boldsymbol{x}(k). \tag{4.41}$$

Moreover, the GUAS of the closed-loop switched polytopic system (3.2) is preserved provided that for each $\sigma_k \in \hat{\mathbb{S}}_p$ it holds that

$$\sum_{\hat{\sigma}\in\hat{\mathbb{S}}_p} \xi_{1\hat{\sigma}} = 1 \quad \text{and} \quad \boldsymbol{P}_{\sigma_k}(1) \succeq \sum_{\hat{\sigma}\in\hat{\mathbb{S}}_p} \xi_{1\hat{\sigma}}\boldsymbol{P}_{\hat{\sigma}}(0) \tag{4.42}$$

with the non-negative scalars $\xi_{1\hat{\sigma}} \in \mathbb{R}_0^+$.

Proof. The proof follows the same line as the proof of Theorem 4.3 except the fact that a one-step cyclic version of each $\sigma_k \in \hat{\mathbb{S}}_p$ is not necessarily contained in the relaxed set $\hat{\mathbb{S}}_p$. Thus, inequality (4.38) is not valid anymore. However, from (4.42) and (4.32) we obtain

$$\boldsymbol{x}^T(1)\boldsymbol{P}_{\overline{\sigma}_0^*}(0)\boldsymbol{x}(1) \geq \sum_{\hat{\sigma}\in\hat{\mathbb{S}}_p} \xi_{1\hat{\sigma}}\boldsymbol{x}^T(1)\boldsymbol{P}_{\hat{\sigma}}(0)\boldsymbol{x}(1). \tag{4.43}$$

Using the relationship introduced in [GC06b, GC06a] that

$$\begin{aligned}\min_{\hat{\sigma}\in\hat{\mathbb{S}}_p} \boldsymbol{x}^T(k)\boldsymbol{P}_{\hat{\sigma}}(0)\boldsymbol{x}(k) &= \min_{\xi_{1\hat{\sigma}}} \sum_{\hat{\sigma}\in\hat{\mathbb{S}}_p} \xi_{1\hat{\sigma}}\boldsymbol{x}^T(k)\boldsymbol{P}_{\hat{\sigma}}(0)\boldsymbol{x}(k)\\ &\leq \sum_{\hat{\sigma}\in\hat{\mathbb{S}}_p} \xi_{1\hat{\sigma}}\boldsymbol{x}^T(k)\boldsymbol{P}_{\hat{\sigma}}(0)\boldsymbol{x}(k),\end{aligned} \tag{4.44}$$

inequality (4.43) implies then that

$$\boldsymbol{x}^T(1)\boldsymbol{P}_{\overline{\sigma}_0^*}(0)\boldsymbol{x}(1) \geq \boldsymbol{x}^T(1)\boldsymbol{P}_{\sigma_1^*}(0)\boldsymbol{x}(1). \tag{4.45}$$

Inequality (4.38) is thus valid again, allowing the rest of the proof of Theorem 4.3 to be followed. This completes the proof □

What missing yet is the determination of the relaxed set $\hat{\mathbb{S}}_p$. To this end, we first introduce a relaxation factor $\alpha \in \mathbb{R}$ with $\alpha \geq 1$. Based on the design parameter α, an

Algorithm 4.1 Relaxed Set $\hat{\mathbb{S}}_p$ Construction

Input: Relaxation factor α, set of p-periodic switching-feedback sequences $\mathbb{S}_p$

Output: Relaxed set of p-periodic switching-feedback sequences $\hat{\mathbb{S}}_p \subset \mathbb{S}_p$

for each $\sigma \in \mathbb{S}_p$ **do**

  if $\nexists\ \xi_{0\hat{\sigma}} \geq 0,\ \sum_{\hat{\sigma} \in \hat{\mathbb{S}}_p} \xi_{0\hat{\sigma}} = 1\text{:}\ \alpha \boldsymbol{P}_\sigma(0) \succeq \sum_{\hat{\sigma} \in \hat{\mathbb{S}}_p} \xi_{0\hat{\sigma}} \boldsymbol{P}_{\hat{\sigma}}(0)$ **then**

    $\hat{\mathbb{S}}_p = \hat{\mathbb{S}}_p \cup \{\sigma\}$ // *add the p-periodic switching-feedback sequence* σ

    if $\nexists\ \xi_{1\hat{\sigma}} \geq 0,\ \sum_{\hat{\sigma} \in \hat{\mathbb{S}}_p} \xi_{1\hat{\sigma}} = 1\text{:}\ \boldsymbol{P}_\sigma(1) \succeq \sum_{\hat{\sigma} \in \hat{\mathbb{S}}_p} \xi_{1\hat{\sigma}} \boldsymbol{P}_{\hat{\sigma}}(0)$ **then**

      $\hat{\mathbb{S}}_p = \hat{\mathbb{S}}_p \cup \{\overline{\sigma}\}$ // *add the one-step cyclic shift of* σ

    end if

  end if

end for

algorithm for constructing the relaxed set $\hat{\mathbb{S}}_p$ while preserving closed-loop stability based on (4.42) is given in Algorithm 4.1. The algorithm is based on retaining some of the p-periodic switching-feedback sequences $\sigma \in \mathbb{S}_p$ in the relaxed set $\hat{\mathbb{S}}_p$ such that

$$\min_{\hat{\sigma} \in \hat{\mathbb{S}}_p} \boldsymbol{x}^T(k) \boldsymbol{P}_{\hat{\sigma}}(0) \boldsymbol{x}(k) \leq \boldsymbol{x}^T(k) \alpha \boldsymbol{P}_\sigma(0) \boldsymbol{x}(k) \tag{4.46}$$

holds $\forall \boldsymbol{x}(k) \in \mathbb{R}^{n+m}$ and $\forall \sigma \in \mathbb{S}_p$. The relaxation condition (4.46) holds if, cf. (4.44), the linear matrix inequality

$$\sum_{\hat{\sigma} \in \hat{\mathbb{S}}_p} \xi_{0\hat{\sigma}} = 1 \quad \text{and} \quad \alpha \boldsymbol{P}_\sigma(0) \succeq \sum_{\hat{\sigma} \in \hat{\mathbb{S}}_p} \xi_{0\hat{\sigma}} \boldsymbol{P}_{\hat{\sigma}}(0) \tag{4.47}$$

holds for each $\sigma \in \mathbb{S}_p$ with the non-negative LMI variables $\xi_{0\hat{\sigma}} \in \mathbb{R}_0^+$. In the case that (4.47) does not hold, the related switching-feedback sequence σ is retained in the relaxed set $\hat{\mathbb{S}}_p$. For each $\hat{\sigma} \in \hat{\mathbb{S}}_p$, the LMI stability preservation condition (4.42) is further evaluated. For those where (4.42) is not feasible, the corresponding one-step cyclically shifted versions $\overline{\hat{\sigma}}$ defined in (4.33) are further retained in the relaxed set $\hat{\mathbb{S}}_p$ as well.

Remark 4.11. For $|\hat{\mathbb{S}}_p|$ reduction, the elements of the set $\mathbb{S}_p$ should be presorted based on the trace of $\boldsymbol{P}_\sigma(0)$ as proposed in [Gör12, Section 4.2.4] before running Algorithm 4.1. Thus the p-periodic switching-feedback sequences $\sigma \in \mathbb{S}_p$ with the presumably minimum upper bound on the worst-case cost function (4.16) are added first to the relaxed set $\hat{\mathbb{S}}_p$ in the case that (4.47) is not feasible.

Remark 4.12. Based on the relaxation condition (4.46), the worst-case cost resulting for the relaxed periodic codesign strategy (4.41) is upper bounded by

$$J_p^{\max}(k) < \min_{\hat{\sigma} \in \hat{\mathbb{S}}_p} \boldsymbol{x}^T(k) \boldsymbol{P}_{\hat{\sigma}}(0) \boldsymbol{x}(k) \leq \min_{\sigma \in \mathbb{S}_p} \boldsymbol{x}^T(k) \alpha \boldsymbol{P}_\sigma(0) \boldsymbol{x}(k) \quad \forall k \in \mathbb{N}_0. \tag{4.48}$$

4.6 Solution based on Optimal Pointer Placement

The periodic codesign strategy (4.41) based on relaxation offers the opportunity to trade-off between the online computational complexity and the resulting control performance through the tuning parameter α. Alternatively, we can offline choose one p-periodic switching-feedback sequence $\sigma_{\text{off}} \in \mathbb{S}_p$ and then online determine the optimal pointer place $\kappa^* \in \{0, 1, \dots, p-1\}$ for the current state $\boldsymbol{x}(k)$ as proposed in [BÇH06]. The relaxed set of p-periodic switching-feedback sequences $\hat{\mathbb{S}}_p$ contains thus only one element, namely σ_{off}. A reasonable choice of σ_{off} can be done either based on the initial value $\boldsymbol{x}(0)$ as

$$\sigma_{\text{off}} = \arg\min_{\sigma \in \mathbb{S}_p} \boldsymbol{x}^T(0)\boldsymbol{P}_\sigma(0)\boldsymbol{x}(0) \tag{4.49}$$

or based on the trace operator as

$$\sigma_{\text{off}} = \arg\min_{\sigma \in \mathbb{S}_p} \text{tr}\big(\boldsymbol{P}_\sigma(0)\big). \tag{4.50}$$

Definition 4.4 *The pointer place κ chosen at time instant t_k is denoted with abuse of notation as κ_k.*

Based on the setup above, Problem 4.2 is then modified as

Problem 4.3 *For the closed-loop switched polytopic system* (3.2) *and the current state $\boldsymbol{x}(k)$ find the optimal pointer place κ_k^* for the p-periodic switching-feedback sequence σ_{off} such that the upper bound on the cost function* (4.12) *is minimized, i.e.*

$$\min_{\kappa \in \{0,1,\dots,p-1\}} \boldsymbol{x}^T(k)\boldsymbol{P}_{\sigma_{\text{off}}}(\kappa)\boldsymbol{x}(k) \qquad \text{subject to (4.13).} \tag{4.51}$$

Apply then the κ_k^*-th element of σ_{off}, i.e. $\sigma_{\text{off}}(\kappa_k^*) = \big(j(\kappa_k^*), \boldsymbol{K}(\kappa_k^*)\big)$, to the closed-loop switched polytopic system (3.2) *and reoptimize at next time instant t_{k+1}.*

Theorem 4.5 *The solution to Problem 4.3 for the current state $\boldsymbol{x}(k)$ and the switching-feedback sequence $\sigma_{\text{off}} \in \mathbb{S}_p$ is given by*

$$\kappa_k^* = \arg \min_{\kappa \{0,1,\dots,p-1\}} \boldsymbol{x}^T(k)\boldsymbol{P}_{\sigma_{\text{off}}}(\kappa)\boldsymbol{x}(k). \tag{4.52}$$

Moreover, the closed-loop switched polytopic system (3.2) *under the optimal pointer placement (OPP) strategy* (4.52) *is GUAS.*

Proof. Since the switching-feedback sequence σ_{off} is an element of the set $\mathbb{S}_p$, the constraints in (4.51) are thus satisfied. Moreover, the optimal pointer place at initial time instant t_0 is given based on (4.52) by

$$\kappa_0^* = \arg \min_{\kappa \{0,1,\dots,p-1\}} \boldsymbol{x}^T(0)\boldsymbol{P}_{\sigma_{\text{off}}}(\kappa)\boldsymbol{x}(0). \tag{4.53}$$

A reoptimization is then performed at next time instant t_1, leading to

$$\kappa_1^* = \arg \min_{\kappa\{0,1,\ldots,p-1\}} \boldsymbol{x}^T(1)\boldsymbol{P}_{\sigma_{\text{off}}}(\kappa)\boldsymbol{x}(1). \tag{4.54}$$

Due to the reoptimization (4.54) at time instant t_1, we conclude that

$$\boldsymbol{x}^T(1)\boldsymbol{P}_{\sigma_{\text{off}}}(\kappa_1^*)\boldsymbol{x}(1) \leq \boldsymbol{x}^T(1)\boldsymbol{P}_{\sigma_{\text{off}}}(\kappa_0^{*+})\boldsymbol{x}(1) \tag{4.55}$$

where

$$\kappa_0^{*+} = (\kappa_0^* + 1) \bmod p. \tag{4.56}$$

From the stability condition (4.13), we further conclude that

$$\boldsymbol{x}^T(1)\boldsymbol{P}_{\sigma_{\text{off}}}(\kappa_0^{*+})\boldsymbol{x}(1) < \boldsymbol{x}^T(0)\boldsymbol{P}_{\sigma_{\text{off}}}(\kappa_0^*)\boldsymbol{x}(0). \tag{4.57}$$

From (4.55) and (4.57), we finally conclude that

$$\boldsymbol{x}^T(1)\boldsymbol{P}_{\sigma_{\text{off}}}(\kappa_1^*)\boldsymbol{x}(1) < \boldsymbol{x}^T(0)\boldsymbol{P}_{\sigma_{\text{off}}}(\kappa_0^*)\boldsymbol{x}(0). \tag{4.58}$$

The same conclusion can be made at each time instant t_k which in turn implies the GUAS of the closed-loop switched polytopic system (3.2), i.e. $\lim_{k\to\infty} \boldsymbol{x}(k) = \boldsymbol{0}$. This completes the proof. □

Remark 4.13. The computational complexity of the OPP strategy (4.52) is determined by the period p. The resulting worst-case cost with the p-periodic switching-feedback sequence σ_{off} chosen according to (4.49) is upper bounded by

$$J_p^{\max}(k) < \boldsymbol{x}^T(k)\boldsymbol{P}_{\sigma_{\text{off}}}(\kappa)\boldsymbol{x}(k) \leq \min_{\sigma\in\mathbb{S}_p} \boldsymbol{x}^T(0)\boldsymbol{P}_\sigma(0)\boldsymbol{x}(0) \quad \forall k \in \mathbb{N}_0 \tag{4.59}$$

where κ is defined according to (4.7).

4.7 Illustrative Example

The aim in this section is to evaluate the periodic control and online scheduling strategy proposed above for networked control of three inverted pendulums with the objective of maintaining them in the upright equilibrium $\varphi_i = 0\,\text{rad}$, $\forall i \in \mathbb{M} = \{1, 2, 3\}$. A schematic diagram of the inverted pendulum mounted on a cart is shown in Figure 4.1.

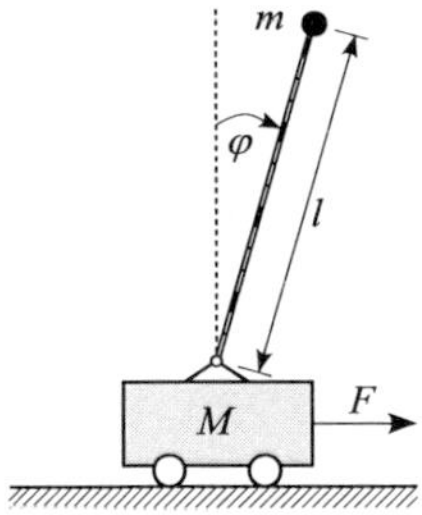

Figure 4.1: Schematic diagram of an inverted pendulum on a cart

Each cart has a mass $M_i = 0.1\,\mathrm{kg}$ and is driven by a motor that exerts a horizontal force F_i in Newton on the cart. The pendulum rods are taken to be massless of lengths

$$l_1 = 0.136\,\mathrm{m}$$
$$l_2 = 0.242\,\mathrm{m}$$
$$l_3 = 0.545\,\mathrm{m}.$$

Moreover, a point mass $m_i = 0.3\,\mathrm{kg}$ is fixed at the tip of each pendulum.

The continuous-time state equation of each inverted pendulum P_i, linearized about the upright equilibrium, is given by

$$\underbrace{\begin{pmatrix} \dot{\varphi}_i(t) \\ \ddot{\varphi}_i(t) \end{pmatrix}}_{\dot{\boldsymbol{x}}_{\mathrm{c}i}(t)} = \underbrace{\begin{pmatrix} 0 & 1 \\ \frac{(m_i+M_i)g}{M_i l_i} & 0 \end{pmatrix}}_{\boldsymbol{A}_{\mathrm{c}i}} \underbrace{\begin{pmatrix} \varphi_i(t) \\ \dot{\varphi}_i(t) \end{pmatrix}}_{\boldsymbol{x}_{\mathrm{c}i}(t)} + \underbrace{\begin{pmatrix} 0 \\ -\frac{1}{M_i l_i} \end{pmatrix}}_{\boldsymbol{b}_{\mathrm{c}i}} \underbrace{F_i(t)}_{u_i(t)} \tag{4.60}$$

where $g = 9.81\,\mathrm{m/s^2}$ is the gravitational acceleration. The derivation of (4.60) can be found e.g. in [KS72, Section 1.2.3]. Each control loop, composed of the plant P_i and its corresponding controller, experiences an uncertain time-varying input delay $\tau_i(k) \in [1\,\mathrm{ms}, 2\,\mathrm{ms}]$ required for transmitting the sensor signal and the control signal. The weighting matrices $\boldsymbol{Q}_{\mathrm{c}i}$ and $R_{\mathrm{c}i}$ of the continuous-time cost function (2.20) are chosen as

$$\boldsymbol{Q}_{\mathrm{c}i} = \begin{pmatrix} 10000 & 0 \\ 0 & 1 \end{pmatrix}, \quad R_{\mathrm{c}i} = 1. \tag{4.61}$$

Since the performance specifications are given in terms of the continuous-time cost function (2.20), a reasonable choice of the sampling interval h_i, $\forall i \in \mathbb{M}$, can be done by designing a continuous-time LQR for each inverted pendulum minimizing (2.20) while neglecting the uncertain time-varying input delay $\tau_i(k)$. Requiring ten samples per rise time of each continuous-time closed-loop system leads to the sampling intervals

$$h_1 = 3\,\mathrm{ms}, \quad h_2 = 4\,\mathrm{ms}, \quad h_3 = 5\,\mathrm{ms}. \tag{4.62}$$

Note that the bounds of the induced input delays given above are not longer than the chosen sampling intervals in (4.62), satisfying thus Assumption 2.2. Following now the modeling procedure in Chapter 2, the discrete-time switched polytopic system (2.18) is obtained with the approximation order $L = 2$. The nominal cost function (2.30) is further obtained with the nominal input delay $\tau_{0i} = 1.5\,\text{ms}$, $\forall i \in \mathbb{M}$.

Based on the resulting NCS model and the associated nominal cost function, Problem 4.2 is first solved for different periods $p \in \{3, \ldots, 10\}$ based on exhaustive search using Theorem 4.3 with the corresponding set of admissible p-periodic switching-feedback sequences $\mathbb{S}_p$ determined based on Theorem 4.2. Problem 4.2 is further solved for different relaxation factors $\alpha \in \{1.00, 1.005, 1.05\}$ based on relaxation using Theorem 4.4 and finally based on optimal pointer placement using Theorem 4.5. The resulting cardinality of the set $\mathbb{S}_p$ and the relaxed set $\hat{\mathbb{S}}_p$ with respect to the period p as well as the computational time required for solving the LMI optimization problem (4.18) in Theorem 4.2 are summarized in Table 4.1.

Period p	$\lvert\mathbb{S}_p\rvert$	$\lvert\hat{\mathbb{S}}_p\rvert$ $\alpha = 1.00$	$\lvert\hat{\mathbb{S}}_p\rvert$ $\alpha = 1.005$	$\lvert\hat{\mathbb{S}}_p\rvert$ $\alpha = 1.05$	Computation Time
3	6	6	6	6	10.51 s
4	36	36	32	12	01.21 min
5	150	150	145	10	04.49 min
6	540	524	458	24	18.98 min
7	1806	1419	924	35	01.24 h
8	5796	3467	1604	72	04.56 h
9	18150	10439	2500	60	16.13 h
10	55980	32308	3688	80	60.55 h

Table 4.1: Offline computational complexity for the PCS strategy

Principally, the higher the chosen period p, the higher the resulting cardinality of $\mathbb{S}_p$ and $\hat{\mathbb{S}}_p$. This is contrasted with the relaxation factor α with which the higher its chosen value, the lower the resulting cardinality of $\hat{\mathbb{S}}_p$. The average (arithmetic mean) of the cost J in (2.29) over one thousand normally distributed random initial states $\boldsymbol{x}_{ci0} = \begin{pmatrix} \varphi_{i0} & \dot{\varphi}_{i0} \end{pmatrix}^T$ with zero expected value and unit covariance matrix is further depicted in Figure 4.2.

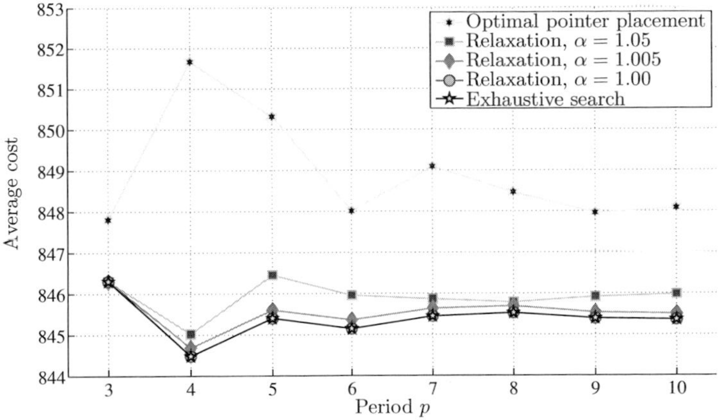

Figure 4.2: Resulting cost versus switching sequence period, averaged over one thousand normally distributed random initial values

As expected, the resulting performance based on exhaustive search (or based on relaxation with $\alpha = 1.0$) is the best. Moreover, the increase of the average cost with the relaxation factor α is quite marginal. On the other hand, a significant decrease of the online complexity w.r.t the factor α is obtained, as seen from the resulting $|\hat{\mathbb{S}}_p|$ in Table 4.1. The resulting performance based on optimal pointer placement with the p-periodic switching-feedback sequence σ_{off} determined by (4.50) is the lowest. A rather more interesting point to be observed is the cost behavior with respect to the period p. Increasing p does not principally provide any benefit from the cost perspective. This is in fact due to the initial value response where all plants must most of the time be simultaneously controlled and hence a 3-periodic switching sequence such as

$$\big(j(0), j(1), j(2)\big) = (1, 2, 3)$$

could be the optimal choice despite the chosen period p. This point is investigated in Chapter 9 in more detail.

For evaluation purposes, the average cost under the well-known TDMA scheduling strategy outlined in Section 1.4 with the p-periodic switching-feedback sequence σ_{off} determined by (4.50) is depicted in Figure 4.3.

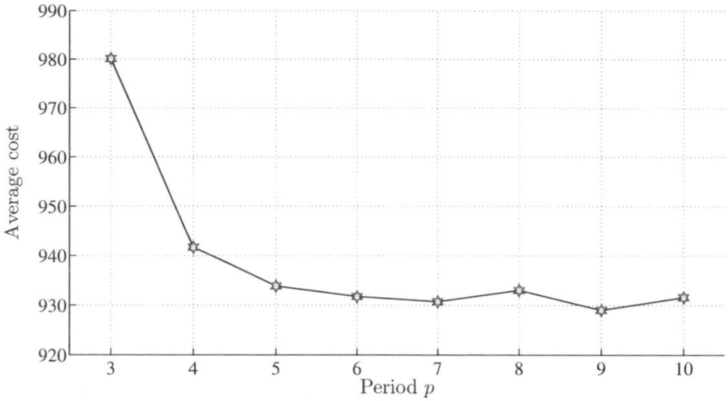

Figure 4.3: Resulting average cost for the TDMA scheduling strategy

It can be observed that the resulting performance is too far away from the performance of the proposed periodic codesign strategy. This is due to its static property where the predefined p-periodic switching-feedback sequence σ_{off} must be followed at runtime despite the system states and demands. Another property to be observed is that the resulting average cost principally decreases with increasing period p since $|\mathbb{S}_{p_2}| > |\mathbb{S}_{p_1}|$ for $p_2 > p_1$ and hence more candidates can be inspected for determining the optimal p-periodic switching-feedback sequence σ_{off}. This property is not always true, as shown in Figure 4.3 between $p = 7$ and $p = 8$, due to the fact that $\mathbb{S}_{p_1} \not\subset \mathbb{S}_{p_2}$ for $p_2 \neq rp_1$, $r = 2, 3, \ldots$. It is thus reasonable to inspect several values for the period p.

Remark 4.14. From the simulation results above, it has been shown that the increase of the period p for an initial value response does not principally provide any benefit from the cost perspective. The situation is however completely different for a disturbance response in which infinite scenarios might occur within a finite simulation time due to the random release instant of the disturbance. The initial value response can thus be seen as a special case of the disturbance response, where all plants are subject to a disturbance impulse (initial value) at the same time instant t_0. The disturbance response under the proposed periodic codesign strategy is studied simulatively and experimentally for a case study in Chapter 9.

5 Receding-Horizon Control and Scheduling (RHCS)

The main codesign problem introduced in Chapter 3 is tackled in Chapter 4 by imposing periodicity on the infinite switching sequence $j(0), \ldots, j(\infty)$ and on the feedback sequence $\boldsymbol{K}(0), \ldots, \boldsymbol{K}(\infty)$. The resulting periodic control and scheduling codesign problem is then solved at each time instant t_k taking the current state $\boldsymbol{x}(k)$ as an initial value $\boldsymbol{x}_0$. In this chapter, alternatively, we decompose the infinite time-horizon cost function J_0 in (2.30) into two parts: J_1 and J_2. In the second part J_2, a stabilizing p-periodic switching-feedback sequence σ defined in (4.1) is considered. Based on this setup, the infinite time-horizon cost function (2.30) is reformulated as a finite time-horizon cost function J_N with the horizon N. Problem 3.1 is then formulated based on the resulting cost function J_N as a receding-horizon control and scheduling codesign problem which can be tackled using dynamic programming. Closed-loop stability under the RHCS strategy is guaranteed under mild assumptions. Both offline and online computational complexity of the RHCS strategy that grow exponentially with N are further adjusted by introducing relaxation in the dynamic programming solution, unlike the PCS strategy where only the online complexity can be adjusted.

5.1 Problem Formulation

Consider the discrete-time quadratic cost function

$$J_0(k) = \sum_{i=0}^{\infty} \boldsymbol{x}^T(k+i)\tilde{\boldsymbol{Q}}_{0j(k+i)}\boldsymbol{x}(k+i) \tag{5.1}$$

where the weighting matrix $\tilde{\boldsymbol{Q}}_{0j(k+i)}$ is defined in (3.5). The cost function (5.1) is decomposed into two parts, i.e.

$$J_0(k) = \underbrace{\sum_{i=0}^{N-1} \boldsymbol{x}^T(k+i)\tilde{\boldsymbol{Q}}_{0j(k+i)}\boldsymbol{x}(k+i)}_{J_1(k)} + \underbrace{\sum_{i=N}^{\infty} \boldsymbol{x}^T(k+i)\tilde{\boldsymbol{Q}}_{0j(k+i)}\boldsymbol{x}(k+i)}_{J_2(k)} \tag{5.2}$$

with the prediction horizon $N \in \mathbb{N}$ as a design parameter.

Lemma 5.1 *The second part $J_2(k)$ in (5.2) is upper bounded by*

$$J_2(k) < \boldsymbol{x}^T(k+N)\boldsymbol{P}_\sigma(0)\boldsymbol{x}(k+N) \tag{5.3}$$

provided that (4.13) is satisfied for a given p-periodic switching-feedback sequence σ.

PROOF. The proof follows the same line as the proof of Theorem 4.1. □

Based on Lemma 5.1, the infinite time-horizon cost function (5.2) can be encapsulated in a finite time-horizon cost function

$$J_N(k) = \boldsymbol{x}^T(k+N)\boldsymbol{P}_\sigma(0)\boldsymbol{x}(k+N) + \sum_{i=0}^{N-1} \boldsymbol{x}^T(k+i)\tilde{\boldsymbol{Q}}_{0j(k+i)}\boldsymbol{x}(k+i). \tag{5.4}$$

The RHCS codesign problem based on the cost function (5.4) can now be formulated as

Problem 5.1 *For the closed-loop switched polytopic system (3.2) and the current state $\boldsymbol{x}(k)$ find the optimal switching sequence $j^*(k), \ldots, j^*(k+N-1)$ and the optimal feedback sequence $\boldsymbol{K}^*(k), \ldots, \boldsymbol{K}^*(k+N-1)$ such that the closed-loop cost function (5.4) is robustly minimized for all $\tau_{j(k)} \in [\underline{\tau}_{j(k)}, \overline{\tau}_{j(k)}]$, i.e.*

$$\min_{\substack{j(k),\ldots,j(k+N-1)\\ \boldsymbol{K}(k),\ldots,\boldsymbol{K}(k+N-1)}} J_N^{\max}(k) \qquad \text{subject to (3.2)} \tag{5.5}$$

where

$$J_N^{\max}(k) \triangleq \max_{\tau_{j(k)},\ldots,\tau_{j(k+N-1)}} J_N(k). \tag{5.6}$$

The optimal switching index $j^*(k)$ and the optimal feedback matrix $\boldsymbol{K}^*(k)$ are then applied to (3.2) at time instant t_k and a reoptimization is done at next time instant t_{k+1}.

Problem 5.1 is solved in the following with some conservatism using dynamic programming [Bel57, Ber05] and using relaxed dynamic programming [LB02, LR06, Ran06]. At each prediction step $i \in \{0, \ldots, N-1\}$, starting from the last one $i = N-1$, the optimal feedback matrix $\boldsymbol{K}^*(k+i)$ is determined by an LMI optimization problem with the objective of minimizing an upper bound on the cost-to-go from $k+i$ to $k+N$. The optimal switching index $j^*(k+i)$ is then determined by explicit enumeration, leading to the switching tree in Figure 5.1. Each branch in the switching tree that results at the prediction step i is indexed by $l(i) \in \mathbb{L}_i$ with the set of indices given by

$$\mathbb{L}_i = \{1, \ldots, M^{N-i}\} \tag{5.7}$$

where M is the number of subsystems and N is the chosen prediction horizon.

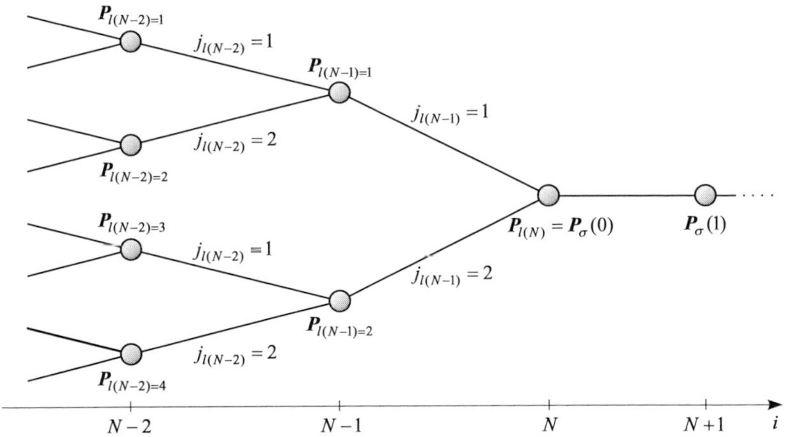

Figure 5.1: Resulting switching tree for $M = 2$

Remark 5.1. From (5.7), one can notice that the number of resulting branches $|\mathbb{L}_i|$ at the prediction step i grows exponentially with the prediction horizon N. This leads of course to some computational problems as discussed in the following.

5.2 Dynamic Programming Solution

Theorem 5.1 *At each prediction step $i \in \{0, \ldots, N-1\}$, consider the optimization problem*

$$\min_{\boldsymbol{K}_{l(i)}} \operatorname{tr}\big(\boldsymbol{P}_{l(i)}\big) \quad \text{subject to} \tag{5.8a}$$

$$\boldsymbol{\mathcal{A}}_{j(i)}^T(i)\boldsymbol{P}_{l(i+1)}\boldsymbol{\mathcal{A}}_{j(i)}(i) - \boldsymbol{P}_{l(i)} \preceq -\tilde{\boldsymbol{Q}}_{0j(i)} \tag{5.8b}$$

for all branch indices $l(i) \in \mathbb{L}_i$ and $l(i+1) \in \mathbb{L}_{i+1}$ of the switching tree in Figure 5.1 with the terminal condition $\boldsymbol{P}_{l(N)} = \boldsymbol{P}_\sigma(0)$. The system matrix $\boldsymbol{\mathcal{A}}_{j(i)}(i)$ is defined in (3.3) *with $\boldsymbol{K}(i) = \boldsymbol{K}_{l(i)}$ and $j(i) = j_{l(i)}$. The minimum upper bound on the objective function* (5.6) *is then given by*

$$J_N^{\max}(k) \leq \boldsymbol{x}^T(k)\boldsymbol{P}_{l^*(0)}\boldsymbol{x}(k) \tag{5.9}$$

with the optimal branch index resulting from

$$l^*(0) = \arg \min_{l(0)\in\mathbb{L}_0} \boldsymbol{x}^T(k)\boldsymbol{P}_{l(0)}\boldsymbol{x}(k). \tag{5.10}$$

The optimal switching index $j^*(k)$ *and feedback matrix* $\boldsymbol{K}^*(k)$ *are given by*

$$\begin{aligned} j^*(k) &= j_{l^*(0)} \\ \boldsymbol{K}^*(k) &= \boldsymbol{K}_{l^*(0)}. \end{aligned} \tag{5.11}$$

PROOF. The proof is based on dynamic programming in which the Bellman equation

$$V^*_{N-i}(i) = \min_{j(i),\, \boldsymbol{K}(i)} \max_{\tau_{j(i)}} \{\boldsymbol{x}^T(i)\tilde{\boldsymbol{Q}}_{0j(i)}\boldsymbol{x}(i) + V^*_{N-(i+1)}(i+1)\} \tag{5.12}$$

with the boundary condition $V^*_0(N) = \boldsymbol{x}^T(N)\boldsymbol{P}_\sigma(0)\boldsymbol{x}(N)$ is used. It is worth to note that the time instant $k = 0$ is assumed within the proof for simplicity of presentation without loss of generality. The value function $V^*_{N-i}(i)$ in (5.12) describes the robust minimum cost-to-go from i to N, where the subindex $N - i$ specifies the steps-to-go.

For the last prediction step $i = N - 1$, the value function is given by

$$V^*_1(N-1) = \min_{\substack{j(N-1) \\ \boldsymbol{K}(N-1)}} \max_{\tau_{j(N-1)}} \{\boldsymbol{x}^T(N-1)\tilde{\boldsymbol{Q}}_{0j(N-1)}\boldsymbol{x}(N-1) + \boldsymbol{x}^T(N)\boldsymbol{P}_\sigma(0)\boldsymbol{x}(N)\}. \tag{5.13}$$

Substituting $\boldsymbol{x}(N) = \boldsymbol{\mathcal{A}}_{j(N-1)}\boldsymbol{x}(N-1)$ into (5.13) yields

$$V^*_1(N-1) = \min_{\substack{j(N-1) \\ \boldsymbol{K}(N-1)}} \max_{\tau_{j(N-1)}} \boldsymbol{x}^T(N-1) \underbrace{\{\tilde{\boldsymbol{Q}}_{0j(N-1)} + \boldsymbol{\mathcal{A}}^T_{j(N-1)}\boldsymbol{P}_{l(N)}\boldsymbol{\mathcal{A}}_{j(N-1)}\}}_{\preceq \boldsymbol{P}_{l(N-1)} \text{ due to (5.8b)}} \boldsymbol{x}(N-1) \tag{5.14}$$

where $\boldsymbol{P}_{l(N)} = \boldsymbol{P}_\sigma(0)$. The minimization of the upper bound

$$\begin{aligned} \boldsymbol{x}^T(N-1)\boldsymbol{P}_{l(N-1)}\boldsymbol{x}(N-1) &\le \lambda_{\max}(\boldsymbol{P}_{l(N-1)})\boldsymbol{x}^T(N-1)\boldsymbol{x}(N-1) \\ &< \operatorname{tr}(\boldsymbol{P}_{l(N-1)})\boldsymbol{x}^T(N-1)\boldsymbol{x}(N-1) \end{aligned} \tag{5.15}$$

with respect to $\boldsymbol{K}(N-1)$ is determined from the optimization problem (5.8). The minimization of (5.15) with respect to $j(N-1) \in \mathbb{M}$ is a combinatorial optimization problem that can be solved by explicit enumeration. This is inherently included in the optimization problem (5.8) via the branch index $l(N-1) \in \mathbb{L}_{N-1} = \{1, \ldots, M\}$. Hence, the minimum upper bound on the value function (5.13) is given by

$$V^*_1(N-1) \le \min_{l(N-1)\in\mathbb{L}_{N-1}} \boldsymbol{x}^T(N-1)\boldsymbol{P}_{l(N-1)}\boldsymbol{x}(N-1). \tag{5.16}$$

For the second to last prediction step $i = N - 2$, the related value function is given by

$$V^*_2(N-2) \le \min_{\substack{j(N-2) \\ \boldsymbol{K}(N-2)}} \max_{\tau_{j(N-2)}} \{\boldsymbol{x}^T(N-2)\tilde{\boldsymbol{Q}}_{0j(N-2)}\boldsymbol{x}(N-2) + \boldsymbol{x}^T(N-1)\boldsymbol{P}_{l(N-1)}\boldsymbol{x}(N-1)\} \tag{5.17}$$

with which the procedure outlined above can again be followed. Iterating backwards until $i = 0$ leads finally to

$$V_N^*(0) \leq \min_{l(0)\in\mathbb{L}_0} \boldsymbol{x}^T(0)\boldsymbol{P}_{l(0)}\boldsymbol{x}(0) \tag{5.18}$$

which in turn implies that

$$J_N^{\max}(0) \leq \min_{l(0)\in\mathbb{L}_0} \boldsymbol{x}^T(0)\boldsymbol{P}_{l(0)}\boldsymbol{x}(0). \tag{5.19}$$

Note that the matrices $\boldsymbol{P}_{l(i)}$ for all $l(i) \in \mathbb{L}_i$ and $i \in \{0, \ldots, N-1\}$ are determined from the optimization problem (5.8) independently of the system state. For $k \neq 0$, the minimum upper bound on the objective function (5.6) is thus given by adapting (5.19) for the current state $\boldsymbol{x}(k)$, i.e.

$$J_N^{\max}(k) \leq \min_{l(0)\in\mathbb{L}_0} \boldsymbol{x}^T(k)\boldsymbol{P}_{l(0)}\boldsymbol{x}(k). \tag{5.20}$$

This completes the proof. □

Remark 5.2. Receding-horizon control and scheduling based on dynamic programming is first introduced in [CA06] using infinite prediction horizon, i.e. $N = \infty$. It has been extended for finite prediction horizon in [GIL11] where closed-loop stability is also addressed. Both of them have however considered switched linear systems (1.5). The dynamic programming solution is generalized here for receding-horizon control and scheduling of the switched polytopic system (3.2), solving the minimax problem (5.5). Moreover, closed-loop stability is also addressed and treated a priori in a unique way.

Lemma 5.2 *The optimization problem* (5.8) *can equivalently be formulated as an LMI optimization problem*

$$\min_{\boldsymbol{G}_{l(i)},\, \boldsymbol{W}_{l(i)},\, \boldsymbol{Z}_{l(i)},\, \beta_{l(i)\ell}^{l(i+1)}} \operatorname{tr}(\boldsymbol{P}_{l(i)}) \quad \text{subject to} \tag{5.21a}$$

$$\begin{pmatrix} \boldsymbol{G}_{l(i)}^T + \boldsymbol{G}_{l(i)} - \boldsymbol{Z}_{l(i)} & * & * & * \\ \boldsymbol{F}_{\mathrm{a}}\boldsymbol{G}_{l(i)} + \boldsymbol{F}_{\mathrm{b}}\boldsymbol{W}_{l(i)} & \beta_{l(i)\ell}^{l(i+1)}\boldsymbol{I} & * & * \\ \sqrt{\boldsymbol{Q}_{0j(i)}}\begin{pmatrix} \boldsymbol{G}_{l(i)} \\ \boldsymbol{W}_{l(i)} \end{pmatrix} & \boldsymbol{0} & \boldsymbol{I} & * \\ \boldsymbol{A}_{j(i)\ell}\boldsymbol{G}_{l(i)} + \boldsymbol{B}_{j(i)\ell}\boldsymbol{W}_{l(i)} & \boldsymbol{0} & \boldsymbol{0} & \boldsymbol{Z}_{l(i+1)} - \beta_{l(i)\ell}^{l(i+1)}\boldsymbol{D}_{j(i)}\boldsymbol{D}_{j(i)}^T \end{pmatrix} \succ \boldsymbol{0} \tag{5.21b}$$

for all $l(i) \in \mathbb{L}_i$, $l(i+1) \in \mathbb{L}_{i+1}$, *and* $\ell \in \{1, \ldots, L+1\}$. *The corresponding feedback matrices* $\boldsymbol{K}_{l(i)}$ *and performance matrices* $\boldsymbol{P}_{l(i)}$ *are given by*

$$\boldsymbol{K}_{l(i)} = \boldsymbol{W}_{l(i)}\boldsymbol{G}_{l(i)}^{-1} \tag{5.22a}$$

$$\boldsymbol{P}_{l(i)} = \boldsymbol{Z}_{l(i)}^{-1}. \tag{5.22b}$$

Proof. Follows the same lines as the proof of Theorem 4.2. □

Remark 5.3. The LMI optimization problem (5.21) can be solved completely offline, reducing the online computation at each time instant t_k to the region membership test (5.10). The offline complexity of solving the LMI optimization problem (5.8b) and the online complexity of evaluating the region membership test (5.10) are both characterized by $|\mathbb{L}_0| = M^N$ that grows exponentially with the prediction horizon N. Such complexity can become computationally intractable (offline) or prohibitive (online) in some practical applications. A relaxation of the dynamic programming solution is therefore considered in the following section to cope with this computational complexity.

GUAS of the closed-loop switched polytopic system (3.2) under the dynamic programming solution (5.11) is, as in classical receding-horizon control, not guaranteed inherently. In the next theorem, we prove the GUAS of (3.2) under (5.11) by augmenting the set $\mathbb{L}_0$ with the branch indices $l(N), \ldots, l(N+p-1)$, where p is the period of the chosen switching-feedback sequence σ in (5.3).

Theorem 5.2 *The closed-loop switched polytopic system* (3.2) *under the dynamic programming solution* (5.11) *is GUAS provided that the augmented set*

$$\tilde{\mathbb{L}}_0 = \mathbb{L}_0 \cup \{l(N), l(N+1), \ldots, l(N+p-1)\} \tag{5.23}$$

is considered in the region membership test (5.10) *instead of* $\mathbb{L}_0$.

Proof. Replacing the set $\mathbb{L}_0$ in (5.10) by the augmented set $\tilde{\mathbb{L}}_0$, the minimum upper bound on the objective function (5.6) at time instant t_0 is given by

$$\min_{l(0)\in\tilde{\mathbb{L}}_0} \boldsymbol{x}^T(0)\boldsymbol{P}_{l(0)}\boldsymbol{x}(0) \leq \boldsymbol{x}^T(0)\boldsymbol{P}_\sigma(0)\boldsymbol{x}(0). \tag{5.24}$$

A reoptimization is then performed at next time instant t_1, leading to the minimum upper bound

$$\min_{l(0)\in\tilde{\mathbb{L}}_0} \boldsymbol{x}^T(1)\boldsymbol{P}_{l(0)}\boldsymbol{x}(1) \leq \boldsymbol{x}^T(1)\boldsymbol{P}_\sigma(1)\boldsymbol{x}(1). \tag{5.25}$$

From the stability condition (4.13), it holds that

$$\boldsymbol{x}^T(k+1)\boldsymbol{P}_\sigma(\kappa^+)\boldsymbol{x}(k+1) < \boldsymbol{x}^T(k)\boldsymbol{P}_\sigma(\kappa)\boldsymbol{x}(k) \quad \forall k \in \mathbb{N}_0 \tag{5.26}$$

where κ is defined in (4.7) and κ^+ in (4.11). Equation (5.26) implies that the upper bound on the objective function (5.6) is monotonically decreasing. Due to the positive definiteness of the matrices $\boldsymbol{P}_\sigma(\kappa)$, $\forall \kappa \in \{0, \ldots, p-1\}$, it holds further that $\lim_{k\to\infty} \boldsymbol{x}(k) = \boldsymbol{0}$. This completes the proof. □

Corollary 5.1 *If for any* $\kappa \in \{0, 1, \ldots, p-1\}$ *there exist non-negative scalars* $\xi_{l(0)} \in \mathbb{R}_0^+$ *with* $l(0) \in \mathbb{L}_0$ *such that*

$$\sum_{l(0)\in\mathbb{L}_0} \xi_{l(0)} = 1 \quad \text{and} \quad \boldsymbol{P}_\sigma(\kappa) \succeq \sum_{l(0)\in\mathbb{L}_0} \xi_{l(0)}\boldsymbol{P}_{l(0)} \tag{5.27}$$

then the set $\tilde{\mathbb{L}}_0 \backslash \{l(N+\kappa)\}$ can be considered in the region membership test (5.10) *while preserving the GUAS of the closed-loop switched polytopic system* (3.2).

PROOF. Satisfaction of (5.27) implies, keeping (4.44) in mind, that

$$\min_{l(0)\in\mathbb{L}_0} \boldsymbol{x}^T(k)\boldsymbol{P}_{l(0)}\boldsymbol{x}(k) \leq \boldsymbol{x}^T(k)\boldsymbol{P}_\sigma(\kappa)\boldsymbol{x}(k) \tag{5.28}$$

for all $\boldsymbol{x}(k) \in \mathbb{R}^{n+m}$. That is, there exists at least one $\boldsymbol{P}_{l(0)}$, $l(0) \in \mathbb{L}_0$, which dominates the matrix $\boldsymbol{P}_\sigma(\kappa)$ for each state $\boldsymbol{x}(k) \in \mathbb{R}^{n+m}$ and hence there is no need to add $\boldsymbol{P}_\sigma(\kappa)$ to the set $\mathbb{L}_0$ for the fulfillment of (5.24) or (5.25) in the proof of Theorem 5.2. □

5.3 Relaxed Dynamic Programming Solution

The aim in this section is to reduce both the offline and the online complexity of the dynamic programming solution, cf. Remark 5.3, at the expense of relaxing the upper bound on the objective function (5.9). More precisely, the upper bound (5.9) is relaxed such that

$$J_N^{\max}(k) \leq \min_{\hat{l}(0)\in\hat{\mathbb{L}}_0} \boldsymbol{x}^T(k)\boldsymbol{P}_{\hat{l}(0)}\boldsymbol{x}(k) \leq \min_{l(0)\in\mathbb{L}_0} \boldsymbol{x}^T(k)\alpha\boldsymbol{P}_{l(0)}\boldsymbol{x}(k) \tag{5.29}$$

with the relaxation factor $\alpha \geq 1$ and the relaxed set $\hat{\mathbb{L}}_0 \subset \mathbb{L}_0$. Two scenarios can be considered to achieve this goal. In the first scenario, the LMI optimization problem (5.21) is first solved for all $i \in \{0,\ldots,N-1\}$ and $l(i) \in \mathbb{L}_i$. The relaxed set $\hat{\mathbb{L}}_0$ is then extracted following the relaxation procedure outlined in Section 4.5. Although the online complexity of evaluating the region membership test (5.10) after replacing the set $\mathbb{L}_0$ by the relaxed set $\hat{\mathbb{L}}_0$ is reduced, this scenario keeps the offline complexity of solving the LMI optimization problem (5.21) unchanged. In the second scenario, the relaxation procedure is done during the dynamic programming solution and not thereafter. This is the main idea of relaxed dynamic programming in which a relaxed set $\hat{\mathbb{L}}_i \subset \mathbb{L}_i$ is extracted at each prediction step i such that (5.29) is finally satisfied. Thus, the offline complexity and the online complexity are both reduced.

Regarding the second scenario, the Bellman equation (5.12) is replaced by the relaxed Bellman equation

$$\underline{V}_{N-i}(i) \leq V_{N-i}(i) \leq \overline{V}_{N-i}(i) \tag{5.30}$$

with the lower bound

$$\underline{V}_{N-i}(i) = \min_{j(i),\, \boldsymbol{K}(i)} \max_{\tau_{j(i)}} \{\boldsymbol{x}^T(i)\tilde{\boldsymbol{Q}}_{0j(i)}\boldsymbol{x}(i) + V_{N-(i+1)}(i+1)\}, \tag{5.31}$$

the upper bound

$$\overline{V}_{N-i}(i) = \min_{j(i),\, \boldsymbol{K}(i)} \max_{\tau_{j(i)}} \{\boldsymbol{x}^T(i)\alpha\tilde{\boldsymbol{Q}}_{0j(i)}\boldsymbol{x}(i) + V_{N-(i+1)}(i+1)\}, \tag{5.32}$$

and the terminal condition

$$V_0(N) = \boldsymbol{x}^T(N)\boldsymbol{P}_\sigma(0)\boldsymbol{x}(N). \tag{5.33}$$

Note that the relaxation factor α is only contained in the upper bound (5.32). Thus the relaxed value function $V_{N-i}(i)$ satisfies

$$V^*_{N-i}(i) \leq \underline{V}_{N-i}(i) \leq V_{N-i}(i) \leq \overline{V}_{N-i}(i) \leq \alpha V^*_{N-i}(i) \tag{5.34}$$

with the optimal value function $V^*_{N-i}(i)$ defined according to (5.12). To fulfill (5.29), the relaxed value function $V_{N-i}(i)$ is parametrized as

$$V_{N-i}(i) = \min_{\hat{l}(i)\in\hat{\mathbb{L}}_i} \boldsymbol{x}^T(i)\boldsymbol{P}_{\hat{l}(i)}\boldsymbol{x}(i). \tag{5.35}$$

The matrices $\boldsymbol{P}_{\hat{l}(i)}$, $\hat{l}(i) \in \hat{\mathbb{L}}_i$, are determined such that the relaxed Bellman equation (5.30) is satisfied with $|\hat{\mathbb{L}}_i|$ as small as possible. Corresponding Theorem 5.1, the lower and the upper bound of the relaxed value function satisfy

$$\underline{V}_{N-i}(i) \leq \min_{l(i)\in\mathbb{L}_i} \boldsymbol{x}^T(i)\underline{\boldsymbol{P}}_{l(i)}\boldsymbol{x}(i) \tag{5.36a}$$

$$\overline{V}_{N-i}(i) \leq \min_{l(i)\in\mathbb{L}_i} \boldsymbol{x}^T(i)\overline{\boldsymbol{P}}_{l(i)}\boldsymbol{x}(i) \tag{5.36b}$$

with the matrices $\underline{\boldsymbol{P}}_{l(i)}$ and $\overline{\boldsymbol{P}}_{l(i)}$ resulting from the optimization problems

$$\min_{\underline{\boldsymbol{K}}_{l(i)}} \operatorname{tr}\big(\underline{\boldsymbol{P}}_{l(i)}\big) \quad \text{subject to} \quad \boldsymbol{\mathcal{A}}^T_{j(i)}(i)\boldsymbol{P}_{\hat{l}(i+1)}\boldsymbol{\mathcal{A}}_{j(i)}(i) - \underline{\boldsymbol{P}}_{l(i)} \preceq -\tilde{\boldsymbol{Q}}_{0j(i)} \tag{5.37a}$$

$$\min_{\overline{\boldsymbol{K}}_{l(i)}} \operatorname{tr}\big(\overline{\boldsymbol{P}}_{l(i)}\big) \quad \text{subject to} \quad \boldsymbol{\mathcal{A}}^T_{j(i)}(i)\boldsymbol{P}_{\hat{l}(i+1)}\boldsymbol{\mathcal{A}}_{j(i)}(i) - \overline{\boldsymbol{P}}_{l(i)} \preceq -\alpha\tilde{\boldsymbol{Q}}_{0j(i)} \tag{5.37b}$$

for all $l(i) \in \mathbb{L}_i$ and $\hat{l}(i+1) \in \hat{\mathbb{L}}_{i+1}$.

Remark 5.4. From the optimization problems in (5.37), only the upper bounds in (5.36) can be obtained rather than the optimal values. Thus, the relaxed Bellman equation (5.30) can not generally be satisfied. Instead, the inequality

$$\min_{l(i)\in\mathbb{L}_i} \operatorname{tr}\big(\underline{\boldsymbol{P}}_{l(i)}\big) \leq \min_{\hat{l}(i)\in\hat{\mathbb{L}}_i} \operatorname{tr}\big(\boldsymbol{P}_{\hat{l}(i)}\big) \leq \min_{l(i)\in\mathbb{L}_i} \operatorname{tr}\big(\overline{\boldsymbol{P}}_{l(i)}\big) \tag{5.38}$$

can be satisfied since the upper bound solutions $\overline{\boldsymbol{P}}_{l(i)}$ of (5.37b) are feasible solutions to the optimization problem (5.37a).

The matrices $\boldsymbol{P}_{\hat{l}(i)}$ can now be obtained while satisfying (5.38) by using some of the lower bound solutions $\underline{\boldsymbol{P}}_{l(i)}$ as the matrices $\boldsymbol{P}_{\hat{l}(i)}$, i.e. $\boldsymbol{P}_{\hat{l}(i)} \in \{\underline{\boldsymbol{P}}_{l(i)}, l(i) \in \mathbb{L}_i\}$. Hence, the first inequality in (5.38) is satisfied. The satisfaction of the second inequality in (5.38) can be implied if it holds for all $l(i) \in \mathbb{L}_i$ that

$$\min_{\hat{l}(i)\in\hat{\mathbb{L}}_i} \operatorname{tr}\big(\boldsymbol{P}_{\hat{l}(i)}\big) \leq \operatorname{tr}\big(\overline{\boldsymbol{P}}_{l(i)}\big). \tag{5.39}$$

Lemma 5.3 *Equation* (5.39) *is implied if there exist non-negative scalars* $\xi_{\hat{l}(i)} \in \mathbb{R}_0^+$ *such that*

$$\sum_{\hat{l}(i)\in\hat{\mathbb{L}}_i} \xi_{\hat{l}(i)} = 1 \quad \text{and} \quad \overline{\boldsymbol{P}}_{l(i)} \succeq \sum_{\hat{l}(i)\in\hat{\mathbb{L}}_i} \xi_{\hat{l}(i)} \boldsymbol{P}_{\hat{l}(i)} \tag{5.40}$$

holds for all $l(i) \in \mathbb{L}_i$.

PROOF. By construction while keeping (4.44) in mind. □

Based on the relaxation criterion (5.40), an algorithm for constructing the relaxed set $\hat{\mathbb{L}}_i$ is given in Algorithm 5.1.

Algorithm 5.1 Relaxed Set $\hat{\mathbb{L}}_i$ Construction

Input: $\underline{\boldsymbol{P}}_{l(i)}, \overline{\boldsymbol{P}}_{l(i)}, \forall l(i) \in \mathbb{L}_i$ // *possibly sorted as in* (5.41)
Output: $\boldsymbol{P}_{\hat{l}(i)}, \forall \hat{l}(i) \in \hat{\mathbb{L}}_i$
  for each $l(i) \in \mathbb{L}_i$ **do**
    if $\nexists\, \xi_{\hat{l}(i)} \geq 0,\ \sum_{\hat{l}(i)\in\hat{\mathbb{L}}_i} \xi_{\hat{l}(i)} = 1$: $\overline{\boldsymbol{P}}_{l(i)} \succeq \sum_{\hat{l}(i)\in\hat{\mathbb{L}}_i} \xi_{\hat{l}(i)} \boldsymbol{P}_{\hat{l}(i)}$ **then**
      $\hat{\mathbb{L}}_i = \hat{\mathbb{L}}_i \cup \{l(i)\}$ // *add branch index*
      $\boldsymbol{P}_{l(i)} = \underline{\boldsymbol{P}}_{l(i)}$ // *add lower bound solution*
    end if
  end for

The algorithm is based on checking the relaxation criterion (5.40) for each $l(i) \in \mathbb{L}_i$. If (5.40) is not feasible for some $l(i)$, the related lower bound solution $\underline{\boldsymbol{P}}_{l(i)}$ is added to the relaxed set $\hat{\mathbb{L}}_i$ since $\mathrm{tr}(\underline{\boldsymbol{P}}_{l(i)}) \leq \mathrm{tr}(\overline{\boldsymbol{P}}_{l(i)})$, cf. Remark 5.4. In order to obtain a relaxed set $\hat{\mathbb{L}}_i$ with $|\hat{\mathbb{L}}_i|$ as small as possible, the lower bound solutions $\underline{\boldsymbol{P}}_{l(i)}$ with lower $\mathrm{tr}(\underline{\boldsymbol{P}}_{l(i)})$ should be added first to the relaxed set. This can be achieved by reassigning the branch indices $l(i) \in \mathbb{L}_i$ of $\underline{\boldsymbol{P}}_{l(i)}, \overline{\boldsymbol{P}}_{l(i)}$ such that

$$\mathrm{tr}(\underline{\boldsymbol{P}}_{l(1)}) \leq \mathrm{tr}(\underline{\boldsymbol{P}}_{l(2)}) \leq \cdots . \tag{5.41}$$

So far, the computational complexity both offline and online are reduced while introducing more suboptimality. For the GUAS of the closed-loop switched polytopic system (3.2) under relaxed dynamic programming, the same technique used for dynamic programming can also be used here. This is printed in the following theorem.

Theorem 5.3 *The closed-loop switched polytopic system* (3.2) *under relaxed dynamic programming is GUAS provided that the branch indices*

$$\{l(N), l(N+1), \ldots, l(N+p-1)\} \tag{5.42}$$

are added to the relaxed set $\hat{\mathbb{L}}_0$.

PROOF. Follows the same line as the proof of Theorem 5.2. □

Remark 5.5. Note that for the relaxation factor $\alpha = 1$ no more suboptimality is introduced. The computational complexity can still be reduced as long as some upper bound solutions $\overline{\boldsymbol{P}}_{l(i)}$ satisfying (5.40) exist. In this case, the corresponding lower bound matrices $\underline{\boldsymbol{P}}_{l(i)}$ can be pruned from the switching tree without introducing any further suboptimality. For further discussion on complexity reduction via pruning, see e.g. [Gör12, Section 2.3].

5.4 Illustrative Example

For the purpose of evaluating the effectiveness of the proposed receding-horizon control and scheduling strategy, the active suspension system of a three-wheeled car is considered in the following. A simplified plant model for each wheel is shown schematically in Figure 5.2(a) and described by a continuous-time linear state equation

$$\underbrace{\begin{pmatrix} \dot{z}_i(t) - \dot{z}_{\mathrm{r}i}(t) \\ \ddot{z}_i(t) \end{pmatrix}}_{\dot{\boldsymbol{x}}_{\mathrm{c}i}(t)} = \underbrace{\begin{pmatrix} 0 & 1 \\ -\frac{k_i}{m_i} & -\frac{b_i}{m_i} \end{pmatrix}}_{\boldsymbol{A}_{\mathrm{c}i}} \underbrace{\begin{pmatrix} z_i(t) - z_{\mathrm{r}i}(t) \\ \dot{z}_i(t) \end{pmatrix}}_{\boldsymbol{x}_{\mathrm{c}i}(t)} + \underbrace{\begin{pmatrix} 0 \\ \frac{1}{m_i} \end{pmatrix}}_{\boldsymbol{b}_{\mathrm{c}i}} \underbrace{F_i(t)}_{u_i(t)} + \underbrace{\begin{pmatrix} -1 \\ \frac{b_i}{m_i} \end{pmatrix}}_{\boldsymbol{b}_{\mathrm{d}i}} \underbrace{\dot{z}_{\mathrm{r}i}(t)}_{d_i(t)}$$

where $z_i - z_{\mathrm{r}i}$ corresponds to the suspension deflection in meters, $\dot{z}_i$ is the derivative of the mass displacement, F_i (control input) is the actuator force in Newton, and $\dot{z}_{\mathrm{r}i}$ (impulsive disturbance) is the derivative of the road displacement. All plants have the same sprung mass (car body) $m_i = 320\,\mathrm{kg}$, suspension stiffness (spring) $k_i = 18000\,\mathrm{N/m}$, and suspension damping (shock absorber) $b_i = 1000\,\mathrm{Ns/m}$ for all $i \in \mathbb{M} = \{1, 2, 3\}$.

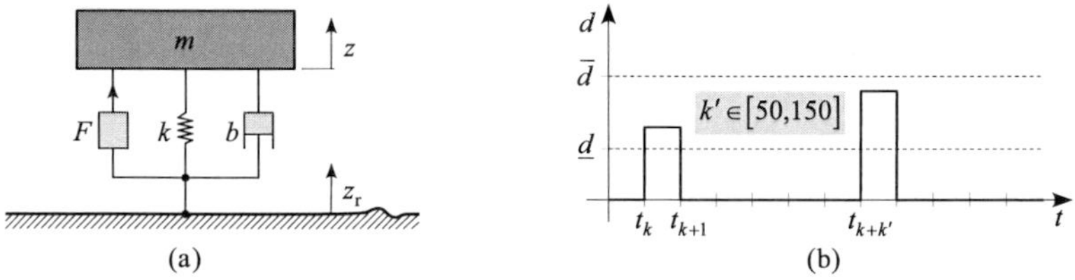

Figure 5.2: (a) A simplified plant model for each wheel and (b) a disturbance model

Consider further a single channel communication network with the uncertain time-varying delays

$$\tau_i^{\mathrm{SC}}(k) \in [1\,\mathrm{ms}, 1.15\,\mathrm{ms}] \tag{5.43a}$$

$$\tau_i^{\mathrm{CA}}(k) \in [1\,\mathrm{ms}, 1.15\,\mathrm{ms}] \tag{5.43b}$$

and the continuous-time cost function (2.20) with the weighting matrices chosen as

$$\boldsymbol{Q}_{ci} = \begin{pmatrix} 10^4 & 0 \\ 0 & 1 \end{pmatrix}, \quad R_{ci} = 10^{-6}$$

for all $i \in \mathbb{M}$. The sampling intervals are further chosen as

$$h_i = h = 5\,\text{ms} \quad \forall i \in \mathbb{M}$$

following the selection procedure outlined in Section 4.7. Finally, the prediction horizon $N = 9$ is chosen.

Based on the given parameters of all plants and the utilized communication network, the discrete-time switched polytopic system (2.18) for the approximation order $L = 5$ is first obtained. The nominal cost function (2.30) is then obtained for the nominal input delay $\tau_{0i} = 2.15\,\text{ms}$, $\forall i \in \mathbb{M}$. Problem 5.1 is then tackled based on dynamic programming using Theorem 5.1. The resulting set $\mathbb{L}_0$ is then replaced by the augmented set $\tilde{\mathbb{L}}_0$ given in (5.23) for stability guarantee. Simulation results for uniformly distributed impulsive disturbance inputs $d_i(t) \in [\underline{d}_i, \overline{d}_i]$, as depicted in Figure 5.2(b) with the lower bound $\underline{d}_i = -10\,\text{m}/\text{sec}$ and the upper bound $\overline{d}_i = 10\,\text{m}/\text{sec}$, and uniformly distributed random input delays $\tau_i(k) \in [2\,\text{ms}, 2.3\,\text{ms}]$ for all $i \in \mathbb{M}$ are shown in Figure 5.3.

Obviously, the network access is granted to the more demanding plant as seen from the output $j(k)$ of the scheduler. For example, within $t_k \in [0.05\,\text{s}, 0.3\,\text{s}]$ in which only P_1 is subject to a disturbance impulse the output of the scheduler $j(k) = 1$ is decided more frequently. The same behavior results for P_2 within $t_k \in [0.3\,\text{s}, 0.55\,\text{s}]$ and for P_3 within $t_k \in [0.55\,\text{s}, 0.75\,\text{s}]$. For $t_k \in [0.8\,\text{s}, 1.0\,\text{s}]$ in which all plants are disturbed, they have granted access alternately. The limited communication resource is thus distributed among the competing plants according to their needs, improving the resulting control performance as shown below.

Problem 5.1 is further tackled by relaxed dynamic programming for different relaxation factors $\alpha \in \{1, 2, \dots, 10\}$ using Algorithm 5.1. The resulting distribution behavior of the limited communication resource is similar to what depicted in Figure 5.3. The resulting cost J defined according to (2.29) within the simulation time $T_{\text{sim}} = 100\,\text{s}$, however, increases with α as shown in Figure 5.4. The resulting cost under the dynamic programming solution (5.11) is also depicted in Figure 5.4: This is the cost for $\alpha = 1$, cf. Remark 5.5. Although the increase of the cost is small, the decrease in $|\hat{\mathbb{L}}_i|$ illustrated in Figure 5.5 is quite significant. Even for $\alpha = 1$ where no cost relaxation exists, the amount of reduction of the online computational complexity due to pruning (measured by $|\hat{\mathbb{L}}_0|$) with respect to the dynamic programming solution is about

$$1 - \frac{|\hat{\mathbb{L}}_0|}{|\mathbb{L}_0|} = 1 - \frac{757}{19683} \cong 96\%.$$

To sum up, the simulation results have shown the capability of the proposed receding-horizon control and scheduling strategy to systematically balance between control performance and offline/online complexity, making it an attractive choice for many networked control applications.

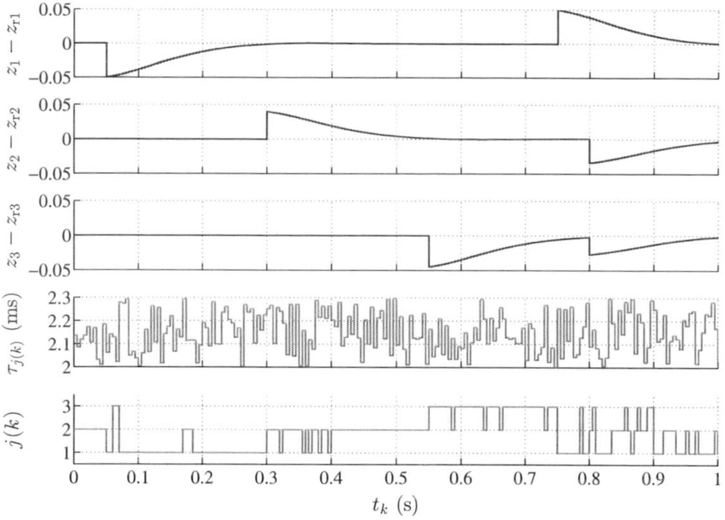

Figure 5.3: Simulation results for the dynamic programming solution under uniformly distributed impulsive disturbances $d_i(t)$ and random input delays $\tau_i(k)$

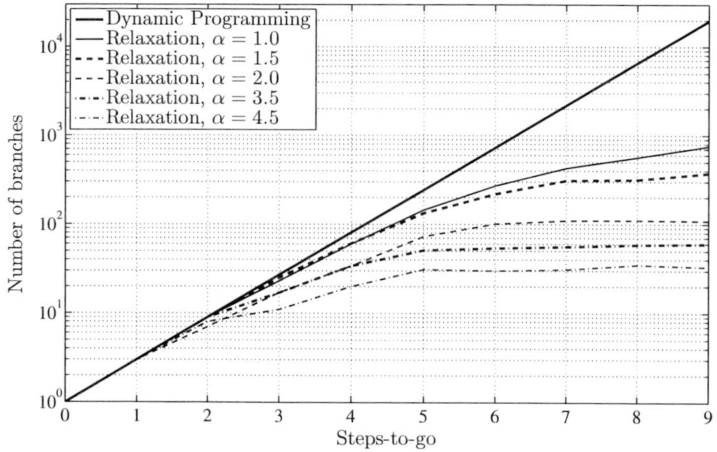

Figure 5.5: Number of branches $|\hat{\mathbb{L}}_i|$ in the resulting switching tree

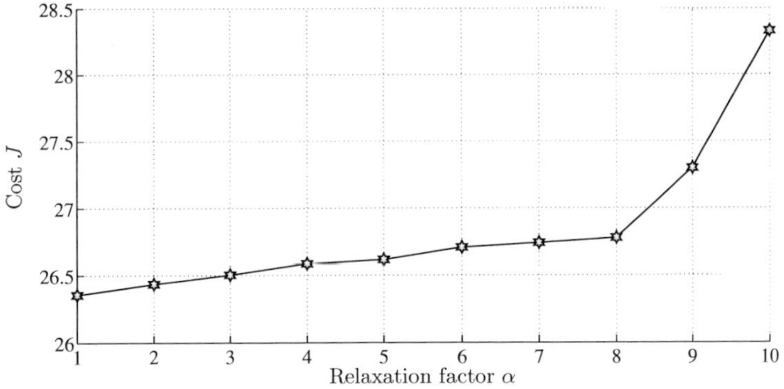

Figure 5.4: Resulting cost under relaxed dynamic programming

Remark 5.6. In the simulation example above, the effect of the relaxation factor α on the resulting control performance and on the resulting complexity has been studied. A more interesting point to be further addressed is the effect of the prediction horizon N on the control performance. This effect has been studied in Chapter 9 for a case study.

6 Implementation-Aware Control and Scheduling (IACS)

Two control and scheduling codesign strategies are proposed in the last two chapters for tackling Problem 3.1, namely the PCS and the RHCS strategy. Within the PCS framework, Problem 3.1 is first transformed into Problem 4.1 and then solved by periodic control and explicit enumeration. The switching index $j(k)$ and the feedback matrix $\boldsymbol{K}(k)$ are determined online based on the region membership test (4.34) or any of its relaxed versions such as (4.41) or (4.52). The same scenario is followed within the RHCS framework in which Problem 3.1 is first transformed into Problem 5.1 and then solved based on (relaxed) dynamic programming. The switching index $j(k)$ and the feedback matrix $\boldsymbol{K}(k)$ are determined online based on the region membership test (5.10).

In both of the codesign strategies above, the feedback matrix $\boldsymbol{K}(k)$ associated to each region is designed offline for a specific switching sequence which is not necessarily the one chosen online and also for all $\boldsymbol{x}(k) \in \mathbb{R}^{n+m}$. In other words, the resulting region membership test (implementation details of the PCS/RHCS strategy) is not considered during the design phase of the feedback matrices. Starting from this aspect, an implementation-aware control and scheduling strategy that considers the implementation details during the codesign process is proposed in this chapter. The feedback matrix $\boldsymbol{K}(k)$ is designed online for the current state $\boldsymbol{x}(k)$ while taking the implementation details of the switching index $j(k)$ into account. Since the proposed codesign problem is solved online, its feasibility is further analyzed and a certificate for closed-loop stability is finally given.

6.1 Problem Formulation

Recall the state feedback control law (3.1) with the feedback matrix $\boldsymbol{K}_i(k)$ replaced by the switched feedback matrix $\boldsymbol{K}_{ij(k)}$, $j(k) \in \mathbb{M}$, i.e.

$$\boldsymbol{u}_i(k) = \boldsymbol{K}_{ij(k)} \boldsymbol{x}_i(k). \tag{6.1}$$

Equation (6.1) implies that the feedback matrices $\boldsymbol{K}_i(k)$ are now restricted to only have M different values. These values must be determined during the codesign process discussed in the following. Substituting the control law (6.1) into the NCS model (2.18) results in a closed-loop switched polytopic system

$$\boldsymbol{x}(k+1) = \underbrace{\left(\sum_{\ell=1}^{L+1} \mu_\ell(\rho_{j(k)}) \tilde{\boldsymbol{A}}_{j(k)\ell} + \boldsymbol{D}_{j(k)} \boldsymbol{E}_{j(k)} \tilde{\boldsymbol{F}}_{j(k)} \right)}_{\boldsymbol{\mathcal{A}}_{j(k)}(k)} \boldsymbol{x}(k) \tag{6.2}$$

where

$$\tilde{\boldsymbol{A}}_{j(k)\ell} = \boldsymbol{A}_{j(k)\ell} + \boldsymbol{B}_{j(k)\ell} \boldsymbol{K}_{j(k)} \tag{6.3a}$$
$$\tilde{\boldsymbol{F}}_{j(k)} = \boldsymbol{F}_{\mathrm{a}} + \boldsymbol{F}_{\mathrm{b}} \boldsymbol{K}_{j(k)} \tag{6.3b}$$
$$\boldsymbol{K}_{j(k)} = \operatorname{diag}\big(\boldsymbol{K}_{1j(k)}, \ldots, \boldsymbol{K}_{Mj(k)}\big). \tag{6.3c}$$

Consider further a discrete-time quadratic cost function

$$J_0(k) = \sum_{i=0}^{\infty} \boldsymbol{x}^T(k+i) \tilde{\boldsymbol{Q}}_{0j(k+i)} \boldsymbol{x}(k+i) \tag{6.4}$$

where

$$\tilde{\boldsymbol{Q}}_{0j(k+i)} = \begin{pmatrix} \boldsymbol{I} \\ \boldsymbol{K}_{j(k+i)} \end{pmatrix}^T \boldsymbol{Q}_{0j(k+i)} \begin{pmatrix} \boldsymbol{I} \\ \boldsymbol{K}_{j(k+i)} \end{pmatrix}. \tag{6.5}$$

and the weighting matrix $\boldsymbol{Q}_{0j(k+i)}$ is defined according to (2.30). Consider finally a state-based switching law

$$j^*(k) = \arg \min_{j(k) \in \mathbb{M}} \boldsymbol{x}^T(k) \boldsymbol{P}_{j(k)} \boldsymbol{x}(k) \tag{6.6}$$

with the symmetric and positive definite scheduling matrices $\boldsymbol{P}_{j(k)}$, $\forall j(k) \in \mathbb{M}$, to be determined. The IACS codesign problem is formulated as

Problem 6.1 *For the closed-loop switched polytopic system* (6.2) *and the current state* $\boldsymbol{x}(k)$ *find the optimal feedback matrices* $\boldsymbol{K}_{j(k)}$ *and the optimal scheduling matrices* $\boldsymbol{P}_{j(k)}$ *for all* $j(k) \in \mathbb{M}$ *such that the closed-loop cost function* (6.4) *is robustly minimized for all* $\tau_{j(k)} \in [\underline{\tau}_{j(k)}, \overline{\tau}_{j(k)}]$, *i.e.*

$$\min_{\substack{\boldsymbol{P}_1,\ldots,\boldsymbol{P}_M \\ \boldsymbol{K}_1,\ldots,\boldsymbol{K}_M}} J_0^{\max}(k) \qquad \text{subject to (6.2) and (6.6)} \tag{6.7}$$

where

$$J_0^{\max}(k) \triangleq \max_{\tau_{j(k)},\ldots,\tau_{j(\infty)}} J_0(k). \tag{6.8}$$

Remark 6.1. Problem 6.1 is computationally intractable, cf. Remark 4.1, and hence an upper bound on the objective function (6.8) is derived in the following and considered as a new objective function for a tractable optimization problem.

6.2 Solution based on Lyapunov-Metzler

Consider a piecewise quadratic Lyapunov function

$$V(k) = \min_{j(k)\in\mathbb{M}} \boldsymbol{x}^T(k)\boldsymbol{P}_{j(k)}\boldsymbol{x}(k) = \min_{\xi_1,\dots,\xi_M} \sum_{j(k)=1}^{M} \xi_{j(k)}\boldsymbol{x}^T(k)\boldsymbol{P}_{j(k)}\boldsymbol{x}(k) \tag{6.9}$$

with the non-negative scalars $\xi_{j(k)} \in \mathbb{R}_0^+$ satisfying $\sum_{j(k)=1}^{M} \xi_{j(k)} = 1$ and the Lyapunov matrices $\boldsymbol{P}_{j(k)}$ defined in (6.6). Note that the chosen Lyapunov function (6.9) explicitly takes the switching law (6.6) into account through the consideration of the minimum among the quadratic forms $\boldsymbol{x}^T(k)\boldsymbol{P}_{j(k)}\boldsymbol{x}(k)$, $\forall j(k) \in \mathbb{M}$, at each time instant t_k. The difference of the Lyapunov function, i.e. $\Delta V(k) \triangleq V(k+1) - V(k)$, along trajectories of the closed-loop switched polytopic system (6.2) is given by

$$\begin{aligned}\Delta V(k) &= \min_{\xi_1,\dots,\xi_M} \sum_{j(k+1)=1}^{M} \xi_{j(k+1)}\boldsymbol{x}^T(k+1)\boldsymbol{P}_{j(k+1)}\boldsymbol{x}(k+1) - \boldsymbol{x}^T(k)\boldsymbol{P}_{j^*(k)}\boldsymbol{x}(k)\\ &\le \sum_{j(k+1)=1}^{M} \xi_{j(k+1)}\boldsymbol{x}^T(k+1)\boldsymbol{P}_{j(k+1)}\boldsymbol{x}(k+1) - \boldsymbol{x}^T(k)\boldsymbol{P}_{j^*(k)}\boldsymbol{x}(k)\end{aligned} \tag{6.10}$$

with the optimal switching index $j^*(k)$ determined from (6.6).

Remark 6.2. The Lyapunov function (6.9) is first introduced in [GC06b, GC06a] for analyzing the stability of autonomous switched systems. It has further been utilized in [DHWH11] for analyzing the stability of switched polytopic systems with a given controller. We use it here for *online* control and scheduling codesign of general switched polytopic systems with additive norm-bounded uncertainty using the switching law (6.6).

Theorem 6.1 *If the difference of the Lyapunov function* (6.10) *satisfies*

$$\Delta V(k) < -\boldsymbol{x}^T(k)\tilde{\boldsymbol{Q}}_{0j(k)}\boldsymbol{x}(k) \qquad \forall \boldsymbol{x}(k) \neq \boldsymbol{0} \tag{6.11}$$

with $\tilde{\boldsymbol{Q}}_{0j(k)}$ *defined in* (6.5), *the objective function* (6.8) *is then upper bounded by*

$$\begin{aligned}J_0^{\max}(k) &< \boldsymbol{x}^T(k)\boldsymbol{P}_{j^*(k)}\boldsymbol{x}(k) && (6.12a)\\ &< \operatorname{tr}\big(\boldsymbol{P}_{j(k)}\big)\boldsymbol{x}^T(k)\boldsymbol{x}(k). && (6.12b)\end{aligned}$$

PROOF. Follows the same line as the proof of Theorem 4.1. □

Remark 6.3. Note that the switching index $j(k) \subset \mathbb{M}$ in (6.12b) does not necessarily have to be equal to the optimal switching index $j^*(k)$ in (6.12a).

Problem 6.1 can now based on the upper bound on the objective function (6.12) be reformulated as

Problem 6.2 *For the closed-loop switched polytopic system (6.2) and the current state $\boldsymbol{x}(k)$ find the optimal feedback matrices $\boldsymbol{K}_{j(k)}$ and the optimal scheduling matrices $\boldsymbol{P}_{j(k)}$ for all $j(k) \in \mathbb{M}$ such that the upper bound on the objective function (6.12) is minimized, i.e.*

$$\min_{\substack{\boldsymbol{P}_1,\ldots,\boldsymbol{P}_M \\ \boldsymbol{K}_1,\ldots,\boldsymbol{K}_M}} \boldsymbol{x}^T(k)\boldsymbol{P}_{j^*(k)}\boldsymbol{x}(k) \qquad \text{subject to (6.11).} \tag{6.13}$$

Problem 6.2 can be expressed as an LMI optimization problem, yet an additional definition is in order.

Definition 6.1 ([GC06b]) *A matrix $\Xi \in \mathbb{R}^{M\times M}$ of non-negative entries $\xi_{j(k)j(k+1)} \in \mathbb{R}_0^+$ such that the elements in each row sum to one, i.e.*

$$\sum_{j(k+1)=1}^{M} \xi_{j(k)j(k+1)} = 1 \qquad \forall j(k) \in \mathbb{M}, \tag{6.14}$$

is called a Metzler matrix.

Theorem 6.2 *Given a Metzler matrix Ξ with non-negative entries $\xi_{j(k)j(k+1)} \in \mathbb{R}_0^+$ satisfying (6.14). The solution to Problem 6.2 results from the LMI optimization problem*

$$\min_{\boldsymbol{G}_{j(k)},\, \boldsymbol{W}_{j(k)},\, \boldsymbol{Z}_{j(k)},\, \gamma,\, \beta_{j(k)\ell}} \gamma \qquad \text{subject to} \tag{6.15a}$$

$$\begin{pmatrix} \gamma & \boldsymbol{x}^T(k) \\ \boldsymbol{x}(k) & \boldsymbol{Z}_{j^*(k)} \end{pmatrix} \succeq \boldsymbol{0} \tag{6.15b}$$

$$\begin{pmatrix} \boldsymbol{G}_{j(k)}^T + \boldsymbol{G}_{j(k)} - \boldsymbol{Z}_{j(k)} & * & * & * & \cdots & * \\ \boldsymbol{F}_\mathrm{a}\boldsymbol{G}_{j(k)} + \boldsymbol{F}_\mathrm{b}\boldsymbol{W}_{j(k)} & \beta_{j(k)\ell}\boldsymbol{I} & * & * & \cdots & * \\ \sqrt{\boldsymbol{Q}_{j(k)}}\begin{pmatrix} \boldsymbol{G}_{j(k)} \\ \boldsymbol{W}_{j(k)} \end{pmatrix} & \boldsymbol{0} & \boldsymbol{I} & * & \cdots & * \\ \boldsymbol{A}_{j(k)\ell}\boldsymbol{G}_{j(k)} + \boldsymbol{B}_{j(k)\ell}\boldsymbol{W}_{j(k)} & \boldsymbol{0} & \boldsymbol{0} & \boldsymbol{X}_{j(k)\ell}^{1} & \cdots & * \\ \vdots & \vdots & \vdots & \vdots & \ddots & \vdots \\ \boldsymbol{A}_{j(k)\ell}\boldsymbol{G}_{j(k)} + \boldsymbol{B}_{j(k)\ell}\boldsymbol{W}_{j(k)} & \boldsymbol{0} & \boldsymbol{0} & -\beta_{j(k)\ell}\boldsymbol{D}_{j(k)}^2 & \cdots & \boldsymbol{X}_{j(k)\ell}^{M} \end{pmatrix} \succ \boldsymbol{0} \tag{6.15c}$$

where

$$\boldsymbol{D}_{j(k)}^2 = \boldsymbol{D}_{j(k)}\boldsymbol{D}_{j(k)}^T \tag{6.16a}$$

$$\boldsymbol{X}_{j(k)\ell}^{j(k+1)} = \xi_{j(k)j(k+1)}^{-1}\boldsymbol{Z}_{j(k+1)} - \beta_{j(k)\ell}\boldsymbol{D}_{j(k)}^2 \quad \forall j(k+1) \in \mathbb{M} \tag{6.16b}$$

for all $\ell \in \{1,\ldots,L+1\}$ and $j(k) \in \mathbb{M}$ with the LMI variables: Block-diagonal matrices $\boldsymbol{G}_{j(k)} \in \mathbb{R}^{(n+m)\times(n+m)}$ and $\boldsymbol{W}_{j(k)} \in \mathbb{R}^{m\times(n+m)}$, symmetric and positive *definite matrices $\boldsymbol{Z}_{j(k)} \in \mathbb{R}^{(n+m)\times(n+m)}$, positive real scalars $\beta_{j(k)\ell} \in \mathbb{R}^+$, and non-negative real scalar $\gamma \in \mathbb{R}_0^+$. The feedback matrices $\boldsymbol{K}_{j(k)}$ and the scheduling matrices $\boldsymbol{P}_{j(k)}$ for all $j(k) \in \mathbb{M}$ are given by*

$$\boldsymbol{K}_{j(k)} = \boldsymbol{W}_{j(k)}\boldsymbol{G}_{j(k)}^{-1} \tag{6.17a}$$

$$\boldsymbol{P}_{j(k)} = \boldsymbol{Z}_{j(k)}^{-1}. \tag{6.17b}$$

PROOF. To prove the implication of the condition (6.11) by the LMI constraints (6.15c), the same line as the proof of Theorem 4.2 can be followed. Furthermore, the objective function in (6.13) can equivalently be written as

$$\min \gamma \qquad \text{subject to} \qquad \boldsymbol{x}^T(k)\boldsymbol{P}_{j^*(k)}\boldsymbol{x}(k) \leq \gamma. \tag{6.18}$$

Applying finally the Schur complement to (6.18) leads equivalently to the LMI constraint (6.15b). This completes the proof. □

Remark 6.4. As a typical example of Metzler matrices, we might choose

$$\Xi = \begin{pmatrix} 0 & 1 & \cdots & 0 \\ \vdots & \vdots & \ddots & \vdots \\ 0 & 0 & \cdots & 1 \\ 1 & 0 & \cdots & 0 \end{pmatrix}. \tag{6.19}$$

Using such a matrix in association with Theorem 6.2 is equivalent to the assumption that there exists an M-periodic switching-feedback sequence σ defined in (4.1) (with the period $p = M$) satisfying the condition (6.11).

Remark 6.5. Since the LMI optimization problem (6.15) is solved online for the current state $\boldsymbol{x}(k)$, a new set of feedback matrices $\boldsymbol{K}_{j(k)}(k)$ and scheduling matrices $\boldsymbol{P}_{j(k)}(k)$, $\forall j(k) \in \mathbb{M}$ are obtained at each time instant t_k. Two problems must therefore be addressed, namely the feasibility of (6.15) for all $\boldsymbol{x}(k) \in \mathbb{R}^{n+m}$ and the stability of the closed-loop switched polytopic system (6.2). These problems are discussed and analyzed in the following two subsections.

The optimal switching index $j^*(k)$ applied to the closed-loop switched polytopic system (6.2) and also required in the LMI optimization problem (6.15) at time instant t_k is determined by evaluating (6.6) based on the scheduling matrices $\boldsymbol{P}_{j(k)}$ that are obtained at last time instant t_{k-1}. The resulting IACS strategy is given in Algorithm 6.1.

Algorithm 6.1 Implementation-Aware Control and Scheduling

Input: Metzler matrix Ξ, feasible scheduling matrices $\boldsymbol{P}_{j(k)}$ $\forall j(k) \in \mathbb{M}$

for each time instant t_k, $k \in \mathbb{N}_0$ **do**
    Measure the current state $\boldsymbol{x}(k)$
    Determine the optimal switching index $j^*(k)$ from (6.6)
    Determine $\boldsymbol{K}_{j(k)}(k)$ and $\boldsymbol{P}_{j(k)}(k)$ from the LMI optimization problem (6.15)
    Apply $j^*(k)$ and $\boldsymbol{K}_{j^*(k)}(k)$ to the switched polytopic system (6.2)
    Store $\boldsymbol{P}_{j(k)}(k)$ for determining $j^*(k+1)$ at next time instant t_{k+1}
end for

The set of feasible scheduling matrices $\boldsymbol{P}_{j(k)}$, $\forall j(k) \in \mathbb{M}$, that are required at initial time instant t_0 for determining $j^*(0)$ can be determined by solving the LMI optimization problem (6.15) taking the upper bound (6.12b) as an objective function instead of (6.12a).

Remark 6.6. Based on Algorithm 6.1, a reoptimization is performed at each time instant t_k such that the upper bound on the objective function (6.12a) is minimized. This recomputation of the feedback matrices $\boldsymbol{K}_{j(k)}$ and the scheduling matrices $\boldsymbol{P}_{j(k)}$, $\forall j(k) \in \mathbb{M}$, can be interpreted as potentially reducing the conservatism in the robust minimization problem (Problem 6.1 or its improved version Problem 6.2).

6.2.1 Feasibility Analysis

In order to prove GUAS of the closed-loop switched polytopic system (6.2), feasibility of the LMI optimization problem (6.15) is first addressed in this subsection.

Lemma 6.1 *The LMI optimization problem* (6.15) *is feasible at each time instant t_k, $k \in \mathbb{N}$, provided that it is feasible at initial time instant t_0.*

Proof. Assume that the LMI optimization problem (6.15) is feasible at t_0, i.e. there exists a factor γ and a set of feedback matrices $\boldsymbol{K}_{j(k)}$ and scheduling matrices $\boldsymbol{P}_{j(k)}$, $\forall j(k) \in \mathbb{M}$, that satisfy the LMI constraints (6.15b) and (6.15c) for the initial state $\boldsymbol{x}(0)$. Since the LMI constraints (6.15c) are state-independent, their feasibility are thus guaranteed for all future time instants t_k, $k \in \mathbb{N}$. On the other hand, the LMI constraint (6.15b) is state-dependent. In order to prove the lemma, therefore, the feasibility of (6.15b) must be guaranteed for all future states $\boldsymbol{x}(k)$, $k \in \mathbb{N}$. More precisely, it is about proving that any feasible solution at time instant t_k is also feasible at time instant t_{k+1}.

Consider the solution γ and $\boldsymbol{P}_{j(k)}$, $\forall j(k) \in \mathbb{M}$, at time instant t_0. it results then from the LMI constraint (6.15b) that

$$\boldsymbol{x}^T(0)\boldsymbol{P}_{j^*(0)}\boldsymbol{x}(0) \leq \gamma. \tag{6.20}$$

Feasibility of the LMI constraint (6.15b) at next time instant t_1 implies that

$$x^T(1)P_{j^*(1)}x(1) \leq \gamma \tag{6.21}$$

holds for all $\tau_{j^*(0)} \in [\underline{\tau}_{j^*(0)}, \overline{\tau}_{j^*(0)}]$ and $x(1) = \mathcal{A}_{j^*(0)}(0)x(0)$ with the optimal switching index $j^*(1)$ determined according to (6.6). Due to the satisfaction of the LMI constraints (6.15c), it holds that

$$\min_{j(1)\in\mathbb{M}} x^T(1)P_{j(1)}x(1) < x^T(0)P_{j^*(0)}x(0) \tag{6.22}$$

for all $\tau_{j^*(0)} \in [\underline{\tau}_{j^*(0)}, \overline{\tau}_{j^*(0)}]$ and $x(1) = \mathcal{A}_{j^*(0)}(0)x(0)$. Taking (6.20) into account, (6.22) corresponds to (6.21) which consequently proves the feasibility of (6.15b) at time instant t_1. The same technique can further be used for proving the feasibility at future time instants t_k, $k > 1$, completing the proof. □

6.2.2 Stability Analysis

Lemma 6.2 *The closed-loop switched polytopic system (6.2) under the IACS strategy given in Algorithm 6.1 is GUAS.*

Proof. Consider the Lyapunov function candidate

$$V^*(k) = x^T(k)P_{j^*(k)}(k)x(k) \tag{6.23}$$

with the time-varying Lyapunov matrix $P_{j^*(k)}(k)$ obtained from the LMI optimization problem (6.15) at time instant t_k. The chosen Lyapunov function (6.23) is positive definite, decrescent, and radially unbounded. What left for proving the GUAS of (6.2) is to guarantee that the Lyapunov candidate (6.23) is strictly decreasing along trajectories of (6.2), i.e.

$$x^T(k+1)P_{j^*(k+1)}(k+1)x(k+1) < x^T(k)P_{j^*(k)}(k)x(k) \tag{6.24}$$

where $x(k+1) = \mathcal{A}_{j^*(k)}(k)x(k)$. Based on the solution of the LMI optimization problem (6.15) at time instant t_k, we obtain that

$$x^T(k+1)P_{j^*(k+1)}(k)x(k+1) < x^T(k)P_{j^*(k)}(k)x(k). \tag{6.25}$$

From Lemma 6.1, it can be concluded that the optimal solution (i.e. $\gamma(k)$, $K_{j(k)}(k)$ and $P_{j(k)}(k)$) obtained at time instant t_k is a feasible solution to (6.15) at next time instant t_{k+1}. Thus, a reoptimization at time instant t_{k+1} must lead to

$$x^T(k+1)P_{j^*(k+1)}(k+1)x(k+1) \leq x^T(k+1)P_{j^*(k+1)}(k)x(k+1) \tag{6.26}$$

which in turn with (6.25) implies (6.24), proving thus the GUAS of (6.2). □

Remark 6.7. In the case that the LMI optimization problem (6.15) with the objective function (6.12b) is considered, the constant factor $x^T(k)x(k)$ does not affect the resulting optimal solution and hence no reoptimization for the current state $x(k)$ is required. Furthermore, the Lyapunov function (6.9) can then be used for analyzing the stability of the closed-loop switched polytopic system (6.2).

6.3 Illustrative Example

Reconsider the setup of the illustrative example outlined in Section 5.4 with the disturbance inputs neglected, i.e. $d_i(t) = 0$ for all $i \in \mathbb{M} = \{1, 2, 3\}$. According to this setup, Problem 6.1 is first solved online based on Algorithm 6.1 (IACS-online) with the Metzler matrix chosen according to (6.19). For evaluation purposes, Problem 6.1 is further solved using Theorem 6.2 but with the upper bound (6.12b) considered as an objective function (IACS-offline). In this case, no reoptimization is required at each time instant t_k since the resulting solution is independent of the state vector $\boldsymbol{x}(k)$. It can thus be solved completely offline.

Simulation results for one hundred realizations of random initial suspension deflection $z_{i0} - z_{\mathrm{r}i0} \in [-0.1\,\mathrm{m}, 0.1\,\mathrm{m}]$ and random input delays $\tau_i(k) \in [2\,\mathrm{ms}, 2.3\,\mathrm{ms}]$ with uniform distribution are depicted in Figure 6.1. The average cost is computed as the sum of the resulting cost J in (2.29) for each realization divided by the number of realizations.

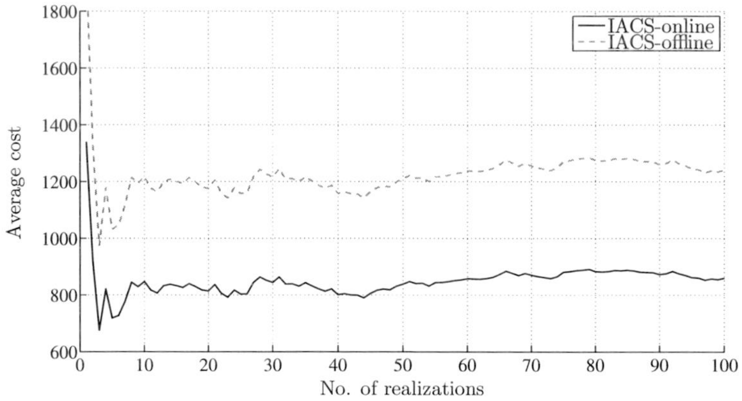

Figure 6.1: Average cost resulting for one hundred realizations of random initial values

As expected, the resulting average cost for the IACS strategy using Algorithm 6.1 is significantly lower than the associated average cost using Theorem 6.2 with the objective function (6.12b). This improvement in the control performance is due to the reoptimization of the control parameters $\boldsymbol{K}_{j(k)}$ and the scheduling parameters $\boldsymbol{P}_{j(k)}$ according to the current state $\boldsymbol{x}(k)$. Thus, the conservatism of the proposed IACS strategy due to the consideration of worst-case scenarios (6.8), cf. Remark 6.6, can significantly be reduced. This reduction is in the sense of getting a faster response compared to the offline counterpart, as depicted in Figure 6.2 for one random realization.

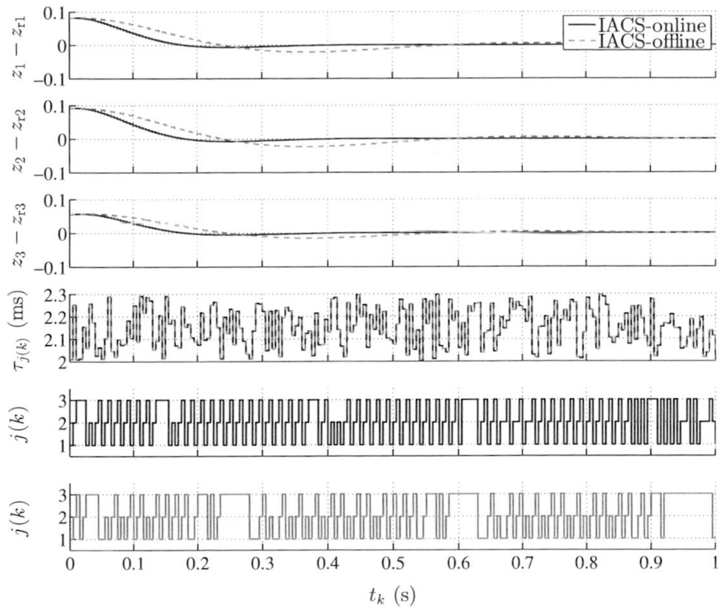

Figure 6.2: Initial-value response under uniformly distributed random input delay $\tau_{j(k)}$

From the initial-value response illustrated in Figure 6.2, it can obviously be seen that the response for the online IACS strategy using Algorithm 6.1 is much faster than the response of the offline counterpart. The slow response for the offline IACS strategy reflects the amount of conservatism we could have if no reoptimization of the control and scheduling parameters is done at each time instant t_k.

7 Event-Based Control and Scheduling (EBCS)

So far, the codesign methods proposed in Chapters 4–6 require transmission of state measurements from all plants to the scheduler for decision making at each sampling instant. This requirement might lead to a large amount of scheduling overhead that subsequently counteract the reactivity property of dynamic schedules to plant demands and/or disturbances. Furthermore, most of the communication networks like Ethernet, WLAN, or CAN are intrinsically event-driven (i.e. utilization is determined by events) rather than time-driven. Hence, the transmission of state measurements in a time-driven fashion may not fit to the network nature (overloading it with unnecessary information). It is thus important to take this phenomenon into account for improving the overall performance. Therefore, an event-based control and scheduling (EBCS) codesign strategy is proposed in this chapter. The main idea of the EBCS strategy consists in assigning an event generator to each plant. Based on these event generators, transmission of new measurements is performed only when necessary. The necessity of transmitting new measurements is decided according to stability and/or performance specifications. Thus, the associated scheduling overhead is significantly reduced while respecting the event-driven nature of the utilized communication network.

7.1 NCS Model Extension

First, the NCS architecture given in Section 2.1 is slightly modified. Each plant P_i is assigned an event generator EG_i for all $i \in \mathbb{M}$. The modified architecture is visualized in Figure 7.1. The structure of the event generator EG_i will be introduced in the following section. To model the modified NCS in Figure 7.1, we recall the solution (2.6) of the continuous-time state equation (2.1) at sampling instants t_k, $k \in \mathbb{N}$, which is given by

$$\boldsymbol{x}_{\mathrm{c}i}(k+1) = \boldsymbol{\Phi}_i(h_{j(k)})\boldsymbol{x}_{\mathrm{c}i}(k) + \boldsymbol{\Gamma}_{1i}(h_{j(k)}, \tau_{ij(k)})\hat{\boldsymbol{u}}_i(k) + \boldsymbol{\Gamma}_{0i}(h_{j(k)}, \tau_{ij(k)})\boldsymbol{u}_i(k) \tag{7.1}$$

with the plant parameters $\boldsymbol{\Phi}_i(t)$, $\boldsymbol{\Gamma}_{1i}(t_1, t_2)$, and $\boldsymbol{\Gamma}_{0i}(t_1, t_2)$ defined according to (2.7). Recall further the state feedback control law (3.1) with the time-varying feedback matrix $\boldsymbol{K}_i(k)$ replaced by the static feedback matrix $\boldsymbol{K}_i$, the augmented state vector $\boldsymbol{x}_i(k)$ by $\boldsymbol{x}_{\mathrm{c}i}(k)$, i.e.

$$\boldsymbol{u}_i(k) = \boldsymbol{K}_i\boldsymbol{x}_{\mathrm{c}i}(k). \tag{7.2}$$

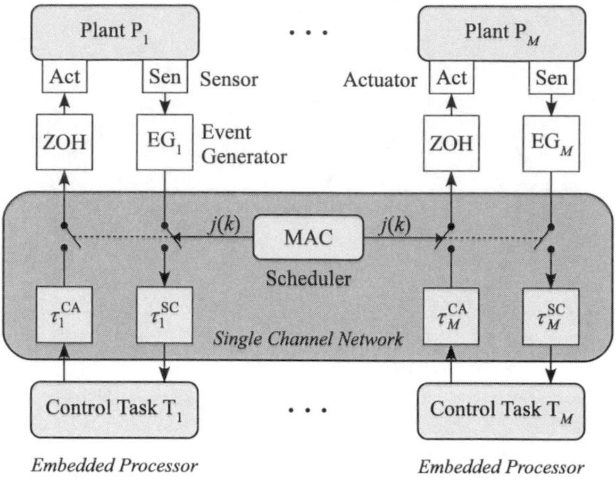

Figure 7.1: Architecture of an event-based networked control system

The last transmitted control vector $\hat{\boldsymbol{u}}_i(k)$ is thus given by

$$\hat{\boldsymbol{u}}_i(k) = \boldsymbol{K}_i \hat{\boldsymbol{x}}_{ci}(k) \tag{7.3}$$

where $\hat{\boldsymbol{x}}_{ci}(k)$ is the last transmitted state measurement of plant P_i. Substituting (7.2) and (7.3) into (7.1) yields a discrete-time closed-loop state equation

$$\begin{pmatrix} \boldsymbol{x}_{ci}(k+1) \\ \hat{\boldsymbol{x}}_{ci}(k+1) \end{pmatrix} = \begin{pmatrix} \boldsymbol{\Phi}_i(h_{j(k)}) + \boldsymbol{\Gamma}_{0i}(h_{j(k)}, \tau_{ij(k)})\boldsymbol{K}_i & \boldsymbol{\Gamma}_{1i}(h_{j(k)}, \tau_{ij(k)})\boldsymbol{K}_i \\ \delta_{ij(k)}\boldsymbol{I} & (1-\delta_{ij(k)})\boldsymbol{I} \end{pmatrix} \begin{pmatrix} \boldsymbol{x}_{ci}(k) \\ \hat{\boldsymbol{x}}_{ci}(k) \end{pmatrix} \tag{7.4}$$

where $\delta_{ij(k)}$ is defined according to (2.5). Equation (7.4) can equivalently be rewritten in terms of the state error vector $\boldsymbol{e}_{ci}(k) = \boldsymbol{x}_{ci}(k) - \hat{\boldsymbol{x}}_{ci}(k)$ as

$$\boldsymbol{x}_i(k+1) = \big(\boldsymbol{A}_{ij(k)} + \boldsymbol{B}_{ij(k)}\tilde{\boldsymbol{K}}_i\big)\boldsymbol{x}_i(k) \tag{7.5}$$

where

$$\boldsymbol{x}_i(k) = \begin{pmatrix} \boldsymbol{x}_{ci}(k) \\ \boldsymbol{e}_{ci}(k) \end{pmatrix}, \quad \tilde{\boldsymbol{K}}_i = \begin{pmatrix} \boldsymbol{K}_i & 0 \\ 0 & \boldsymbol{K}_i \end{pmatrix} \tag{7.6}$$

$$\boldsymbol{A}_{ij(k)} = \begin{pmatrix} \boldsymbol{\Phi}_i(h_{j(k)}) & 0 \\ \boldsymbol{\Phi}_i(h_{j(k)}) - \boldsymbol{I} & (1-\delta_{ij(k)})\boldsymbol{I} \end{pmatrix} \tag{7.7}$$

$$\boldsymbol{B}_{ij(k)} = \begin{pmatrix} \boldsymbol{\Gamma}_i(h_{j(k)}) & -\boldsymbol{\Gamma}_{1i}(h_{j(k)}, \tau_{ij(k)}) \\ \boldsymbol{\Gamma}_i(h_{j(k)}) & -\boldsymbol{\Gamma}_{1i}(h_{j(k)}, \tau_{ij(k)}) \end{pmatrix} \tag{7.8}$$

$$\boldsymbol{\Gamma}_i(h_{j(k)}) = \boldsymbol{\Gamma}_{0i}(h_{j(k)}, \tau_{ij(k)}) + \boldsymbol{\Gamma}_{1i}(h_{j(k)}, \tau_{ij(k)}). \tag{7.9}$$

Following the overapproximation procedure in Section 2.3 results in a closed-loop plant model with polytopic and additive norm-bounded uncertainty

$$\boldsymbol{x}_i(k+1) = \left(\sum_{\ell=1}^{L+1} \mu_\ell(\rho_{j(k)}) \tilde{\boldsymbol{A}}_{ij(k)\ell} + \boldsymbol{D}_{ij(k)} \boldsymbol{E}_{ij(k)} \tilde{\boldsymbol{F}}_i \right) \boldsymbol{x}_i(k) \tag{7.10}$$

where

$$\tilde{\boldsymbol{A}}_{ij(k)\ell} = \boldsymbol{A}_{ij(k)} + \boldsymbol{B}_{ij(k)\ell} \tilde{\boldsymbol{K}}_i \tag{7.11}$$

$$\boldsymbol{D}_{ij(k)} = \begin{pmatrix} \gamma_{ij(k)} \boldsymbol{I} & \gamma_{ij(k)} \boldsymbol{I} \end{pmatrix}^T \tag{7.12}$$

$$\tilde{\boldsymbol{F}}_i = \boldsymbol{F}_i \tilde{\boldsymbol{K}}_i, \quad \boldsymbol{F}_i = \begin{pmatrix} \boldsymbol{0} & \boldsymbol{I} \end{pmatrix}. \tag{7.13}$$

Consequently, the NCS in Figure 7.1 is modeled as a block-diagonal discrete-time switched polytopic system

$$\boldsymbol{x}(k+1) = \underbrace{\left(\sum_{\ell=1}^{L+1} \mu_\ell(\rho_{j(k)}) \tilde{\boldsymbol{A}}_{j(k)\ell} + \boldsymbol{D}_{j(k)} \boldsymbol{E}_{j(k)} \tilde{\boldsymbol{F}} \right)}_{\boldsymbol{\mathcal{A}}_{j(k)}(k)} \boldsymbol{x}(k). \tag{7.14}$$

For the cost function assigned to each plant P_i, we recall further the individual cost function (2.26) with the input delay $\tau_i(k)$ replaced by its nominal value τ_{0i}, i.e.

$$J_{0i} = \sum_{k=0}^{\infty} \begin{pmatrix} \boldsymbol{x}_{\mathrm{c}i}(k) \\ \hat{\boldsymbol{u}}_i(k) \\ \boldsymbol{u}_i(k) \end{pmatrix}^T \boldsymbol{Q}_{0ij(k)} \begin{pmatrix} \boldsymbol{x}_{\mathrm{c}i}(k) \\ \hat{\boldsymbol{u}}_i(k) \\ \boldsymbol{u}_i(k) \end{pmatrix}. \tag{7.15}$$

Substituting (7.2) and (7.3) into (7.15) yields, after reformulating it in terms of the state error vector $\boldsymbol{e}_{\mathrm{c}i}(k)$, a discrete-time closed-loop cost function

$$J_{0i} = \sum_{k=0}^{\infty} \boldsymbol{x}_i^T(k) \tilde{\boldsymbol{Q}}_{0ij(k)} \boldsymbol{x}_i(k) \tag{7.16}$$

where

$$\tilde{\boldsymbol{Q}}_{0ij(k)} = \begin{pmatrix} \boldsymbol{I} & \boldsymbol{0} \\ \boldsymbol{K}_i & -\boldsymbol{K}_i \\ \boldsymbol{K}_i & \boldsymbol{0} \end{pmatrix}^T \boldsymbol{Q}_{0ij(k)} \begin{pmatrix} \boldsymbol{I} & \boldsymbol{0} \\ \boldsymbol{K}_i & -\boldsymbol{K}_i \\ \boldsymbol{K}_i & \boldsymbol{0} \end{pmatrix}. \tag{7.17}$$

The closed-loop cost function assigned to the NCS is thus given by

$$J_0 = \sum_{i=1}^{M} J_{0i} = \sum_{k=0}^{\infty} \boldsymbol{x}^T(k) \tilde{\boldsymbol{Q}}_{0j(k)} \boldsymbol{x}(k). \tag{7.18}$$

7.2 Problem Formulation

Before delving into the details of the event-based codesign problem, we introduce first the structure of the triggering mechanism implemented in the event generator EG_i. Many mechanisms have been proposed in the literature for triggering an event once a necessity exists. These can roughly be classified into three types: threshold-based, cost-based, and rate-based triggering mechanisms. Under threshold-based triggering mechanisms, see e.g. [Tab07,LL10,HDT12], a relative-type threshold (e.g. a function of system's state) or an absolute-type threshold (e.g. a constant scalar) is considered as a design parameter chosen by the control designer. Events are then triggered once the chosen threshold is exceeded, otherwise no reaction is required. The amount of communication reduction is implicitly determined by the chosen thresholds. Under cost-based triggering mechanisms, a physically motivated cost of communication is introduced and combined with the cost function (7.18). Based on the resulting overall cost of control and communication, the optimal triggering mechanism (which can in some cases be posed as threshold-based mechanisms) can be obtained using dynamic programming, see e.g. [XH04, MH09, GPRP13]. Under rate-based mechanisms, finally, a desired rate of communication is explicitly specified at the beginning. The desired rate of communication is considered as an additional constraint to an optimization problem with the objective of minimizing the assigned control cost function (7.18), see e.g. [AHT12, MH12, RSJ13]. Due to the simplicity of integrating them into the codesign procedure given below, threshold-based triggering mechanisms are considered in the following.

Consider the event generator EG_i implementing a discrete-time event-triggering law

$$\sigma_i(k) \triangleq \begin{pmatrix} \boldsymbol{x}_{\mathrm{c}i}(k) \\ \boldsymbol{e}_{\mathrm{c}i}(k) \end{pmatrix}^T \underbrace{\begin{pmatrix} -\boldsymbol{S}_{2i} & \boldsymbol{0} \\ \boldsymbol{0} & \boldsymbol{S}_{1i} \end{pmatrix}}_{\boldsymbol{S}_i} \begin{pmatrix} \boldsymbol{x}_{\mathrm{c}i}(k) \\ \boldsymbol{e}_{\mathrm{c}i}(k) \end{pmatrix} > 0 \tag{7.19}$$

with the weighting matrices $\boldsymbol{S}_{1i} \succ \boldsymbol{0}$ and $\boldsymbol{S}_{2i} \succ \lambda_i \boldsymbol{S}_{1i}$, $\forall i \in \mathbb{M}$, to be determined and the tuning parameters $\lambda_i \in \mathbb{R}_0^+$ for balancing between control performance and network utilization.

Remark 7.1. The event-triggering law (7.19) can equivalently be rewritten as

$$\boldsymbol{e}_{\mathrm{c}i}^T(k)\boldsymbol{S}_{1i}\boldsymbol{e}_{\mathrm{c}i}(k) > \boldsymbol{x}_{\mathrm{c}i}^T(k)\boldsymbol{S}_{2i}\boldsymbol{x}_{\mathrm{c}i}(k) \tag{7.20}$$

where the squared weighted Euclidean norm of the state error $\boldsymbol{e}_{\mathrm{c}i}(k)$ is compared with the squared weighted Euclidean norm of the state $\boldsymbol{x}_{\mathrm{c}i}(k)$ at each time instant t_k. The state-dependent term $\boldsymbol{x}_{\mathrm{c}i}^T(k)\boldsymbol{S}_{2i}\boldsymbol{x}_{\mathrm{c}i}(k)$ serves as a relative threshold that can be tuned via the design parameter λ_i. The higher the chosen value of λ_i, the higher the resulting threshold and usually the lower the utilization of the communication network.

Remark 7.2. Beside the relative threshold in (7.20), we can introduce an absolute threshold $\epsilon_i \in \mathbb{R}_0^+$, leading to the event-triggering law

$$\boldsymbol{e}_{\mathrm{c}i}^T(k)\boldsymbol{S}_{1i}\boldsymbol{e}_{\mathrm{c}i}(k) > \boldsymbol{x}_{\mathrm{c}i}^T(k)\boldsymbol{S}_{2i}\boldsymbol{x}_{\mathrm{c}i}(k) + \epsilon_i \tag{7.21}$$

By introducing ϵ_i, more freedom in adjusting and reducing the number of events can be obtained. GUAS of the closed-loop switched polytopic system (7.14) can not, however, be guaranteed under the event-triggering law (7.21). For more details on extending the proposed work in this chapter for the event-triggering law (7.21) refer to [AGL15a].

Motivated by the proposed event-triggering law (7.19), consider finally the scheduler implementing an event-based switching law

$$j(k) = \begin{cases} 0 & \text{if } \sigma_i(k) \leq 0, \ \forall i \in \mathbb{M} \\ \arg\max_{i \in \mathbb{M}} \sigma_i(k) & \text{otherwise.} \end{cases} \tag{7.22}$$

Remark 7.3. The event-based switching law (7.22) can be seen as an extension of the TOD protocol proposed in [WYB02] with which the most demanding plant is chosen at time instant t_k to be served. Less demanding plants must thus give up and try again at next time instants. In case of the output of the scheduler is $j(k) = 0$, the communication network can be utilized for other non-control tasks T_0 or be idle.

Remark 7.4. In priority-based MAC protocols such as the CAN protocol, the switching law (7.22) can be implemented with a very low scheduling overhead. The main idea is to assign the priority of a node (i.e. plant) dynamically based on the resulting value of $\sigma_i(k)$ at each time instant t_k. The arbitration mechanism utilized in the CAN protocol grants an access to the node with the highest priority in a non-destructive way (i.e. without collision). Therefore, the associated arbitration mechanism "implements" (7.22) without any further hardware and/or software modifications. To avoid the case that two or more nodes have the same value of $\sigma_i(k)$ at time instant t_k, one may additionally prescribe static priorities amongst the nodes.

The event-based codesign problem of the control law (7.2) and the switching law (7.22) can now be formulated as

Problem 7.1 *For the closed-loop switched polytopic system* (7.14) *find the feedback matrices* $\boldsymbol{K}_i$ *in* (7.2) *and the triggering matrices* $\boldsymbol{S}_i$ *in* (7.19) *for all* $i \in \mathbb{M}$ *such that the closed-loop cost function* (7.18) *is robustly minimized for all* $\tau_{j(k)} \in [\underline{\tau}_{j(k)}, \overline{\tau}_{j(k)}]$, *i.e.*

$$\min_{\substack{\boldsymbol{S}_1,\dots,\boldsymbol{S}_M \\ \boldsymbol{K}_1,\dots,\boldsymbol{K}_M}} J_0^{\max} \quad \text{subject to (7.14) and (7.22)} \tag{7.23}$$

where

$$J_0^{\max} \triangleq \max_{\tau_{j(0)},\dots,\tau_{j(\infty)}} J_0. \tag{7.24}$$

Remark 7.5. According to Remark 4.1, Problem 7.1 is computationally intractable and thus an upper bound on the objective function (7.24) is derived in the following and considered as a new objective function for a tractable optimization problem.

7.3 Solution based on the S-Procedure

Consider a switched Lyapunov function

$$V(k) = \boldsymbol{x}^T(k)\boldsymbol{P}_{j(k)}\boldsymbol{x}(k) \tag{7.25}$$

where the Lyapunov matrices $\boldsymbol{P}_{j(k)} \in \mathbb{R}^{(2n\times 2n)}$ are symmetric and positive definite. The difference of the Lyapunov function, i.e. $\Delta V(k) \triangleq V(k+1) - V(k)$, along trajectories of the closed-loop switched polytopic system (7.14) is given by

$$\Delta V(k) = \boldsymbol{x}^T(k)\left(\boldsymbol{\mathcal{A}}^T_{j(k)}(k)\boldsymbol{P}_{j(k+1)}\boldsymbol{\mathcal{A}}_{j(k)}(k) - \boldsymbol{P}_{j(k)}\right)\boldsymbol{x}(k). \tag{7.26}$$

Lemma 7.1 *Based on the event-based switching law (7.22), the state space $\mathbb{R}^{2n}$ is partitioned into regions each expressed by a quadratic form*

$$\mathbb{X}_{j(k)} \triangleq \{\boldsymbol{x}(k) \in \mathbb{R}^{2n} \mid \boldsymbol{x}^T(k)\tilde{\boldsymbol{S}}_{j(k)}\boldsymbol{x}(k) \geq 0\} \tag{7.27}$$

where

$$\tilde{\boldsymbol{S}}_{j(k)} = \begin{cases} \mathrm{diag}\big(-\boldsymbol{S}_1, \ldots, -\boldsymbol{S}_M\big) & \text{for } j(k) = 0 \\ \mathrm{diag}\big(-\boldsymbol{S}_1, \ldots, (M-1)\boldsymbol{S}_{j(k)}, \ldots, -\boldsymbol{S}_M\big) & \text{otherwise.} \end{cases} \tag{7.28}$$

PROOF. To prove (7.27), we consider two scenarios. The first scenario is when at least one event is triggered at time instant t_k, i.e. $j(k) \neq 0$, while the second one when none of the event generators triggers an event, i.e. $j(k) = 0$.

In case of $j(k) \neq 0$, it can be concluded from the event-based switching law (7.22) that

$$\sigma_{j(k)}(k) \geq \sigma_i(k) \qquad \forall i \in \mathbb{M}. \tag{7.29}$$

Equation (7.29) can equivalently be rewritten in terms of the state vector $\boldsymbol{x}(k)$ as

$$\boldsymbol{x}^T(k)\mathrm{diag}(\boldsymbol{0}, \ldots, \boldsymbol{S}_{j(k)}, -\boldsymbol{S}_i, \ldots, \boldsymbol{0})\boldsymbol{x}(k) \geq 0. \tag{7.30}$$

Summing up (7.30) over $i = \{1, \ldots, M\}$ and $i \neq j(k)$ yields the quadratic form given in (7.27) with the block-diagonal matrix $\tilde{\boldsymbol{S}}_{j(k)}$ as defined in (7.28).

In case of $j(k) = 0$, it holds from the event-triggering law (7.19) that

$$\sigma_i(k) \leq 0 \qquad \forall i \in \mathbb{M}. \tag{7.31}$$

Equation (7.31) can also be formulated in the form of (7.27) by following the same line above. This completes the proof. □

Lemma 7.2 *If the difference of the Lyapunov function (7.26) satisfies*

$$\Delta V(k) < -\boldsymbol{x}^T(k)\tilde{\boldsymbol{Q}}_{0j(k)}\boldsymbol{x}(k) \qquad \forall \boldsymbol{x}(k) \in \mathbb{X}_{j(k)} \tag{7.32}$$

with $\tilde{\boldsymbol{Q}}_{0j(k)}$ defined in (7.18), then the objective function (7.24) is upper bounded by

$$J_0^{\max} < \boldsymbol{x}^T(0)\boldsymbol{P}_{j(0)}\boldsymbol{x}(0) \tag{7.33a}$$

$$< \operatorname{tr}\big(\boldsymbol{P}_{j(0)}\big)\boldsymbol{x}^T(0)\boldsymbol{x}(0). \tag{7.33b}$$

PROOF. The proof follows the same line as the proof of Theorem 4.1. The additional restriction, i.e. $\boldsymbol{x}(k) \in \mathbb{X}_{j(k)}$, is due to the partition caused by the event-based switching law (7.22), as described in Lemma 7.1. □

Remark 7.6. Note that the stability condition (7.32) must be satisfied for all combinations of the switching indices $j(k) \in \mathbb{M}_0$ and $j(k+1) \in \mathbb{M}_0$ with the augmented set $\mathbb{M}_0$ given by

$$\mathbb{M}_0 = \mathbb{M} \cup \{0\}. \tag{7.34}$$

Remark 7.7. The satisfaction of the stability condition (7.32) for all combinations of $j(k) \in \mathbb{M}_0$ and $j(k+1) \in \mathbb{M}_0$ implies the existence of quadratic Lyapunov-like functions $V_{j(k)}(k) = \boldsymbol{x}^T(k)\boldsymbol{P}_{j(k)}\boldsymbol{x}(k)$, each assigned to each region $\mathbb{X}_{j(k)}$, that monotonically decrease but only inside their regions. Furthermore, the value of $V_{j(k+1)}(k+1)$ at time instant t_{k+1} is lower than the value of $V_{j(k)}(k)$ at time instant t_k in the case that a switching happens from the region $\mathbb{X}_{j(k)}$ to a region $\mathbb{X}_{j(k+1)}$. Thus, the GUAS of the closed-loop switched polytopic system (7.14) is guaranteed.

The LMI optimization problem corresponding to Problem 7.1 can now be formulated as

Theorem 7.1 *The solution to Problem 7.1 with the objective function (7.24) replaced by the upper bound (7.33b) results from the LMI optimization problem*

$$\min_{\boldsymbol{G},\,\boldsymbol{W},\,\boldsymbol{Z}_{j(k)},\,\boldsymbol{M}_{j(k)},\,\beta_{j(k)\ell}^{j(k+1)}} \operatorname{tr}\big(\boldsymbol{P}_{j(0)}\big) \qquad \text{subject to} \tag{7.35a}$$

$$\begin{pmatrix} \boldsymbol{G}^T + \boldsymbol{G} - \boldsymbol{Z}_{j(k)} - \boldsymbol{M}_{j(k)} & * & * & * \\ \boldsymbol{F}\boldsymbol{W} & \beta_{j(k)\ell}^{j(k+1)}\boldsymbol{I} & * & * \\ \boldsymbol{Q}_{0j(k)}^{1/2}\boldsymbol{X} & 0 & \boldsymbol{I} & * \\ \boldsymbol{A}_{j(k)}\boldsymbol{G} + \boldsymbol{B}_{j(k)\ell}\boldsymbol{W} & 0 & 0 & \boldsymbol{Z}_{j(k+1)} - \beta_{j(k)\ell}^{j(k+1)}\boldsymbol{D}_{j(k)}\boldsymbol{D}_{j(k)}^T \end{pmatrix} \succ 0 \tag{7.35b}$$

where

$$
\begin{aligned}
\boldsymbol{G} &= \mathrm{diag}(\boldsymbol{G}_1, \boldsymbol{G}_1, \ldots, \boldsymbol{G}_M, \boldsymbol{G}_M) && \text{(7.36a)}\\
\boldsymbol{W} &= \mathrm{diag}(\boldsymbol{W}_1, \boldsymbol{W}_1, \ldots, \boldsymbol{W}_M, \boldsymbol{W}_M) && \text{(7.36b)}\\
\boldsymbol{X} &= \mathrm{diag}(\boldsymbol{X}_1, \ldots, \boldsymbol{X}_M) && \text{(7.36c)}\\
\boldsymbol{X}_i &= \begin{pmatrix} \boldsymbol{G}_i & 0 \\ \boldsymbol{W}_i & -\boldsymbol{W}_i \\ \boldsymbol{W}_i & 0 \end{pmatrix} \quad \forall i \in \mathbb{M} && \text{(7.36d)}\\
\tilde{\boldsymbol{M}}_{j(k)} &= \begin{cases} \mathrm{diag}(-\boldsymbol{M}_1, \ldots, -\boldsymbol{M}_M) & \text{for } j(k) = 0 \\ \mathrm{diag}(-\boldsymbol{M}_1, \ldots, (M-1)\boldsymbol{M}_{j(k)}, \ldots, -\boldsymbol{M}_M) & \text{otherwise} \end{cases} && \text{(7.36e)}\\
\boldsymbol{M}_i &= \mathrm{diag}(-\boldsymbol{M}_{2i}, \boldsymbol{M}_{1i}) \quad \forall i \in \mathbb{M} && \text{(7.36f)}\\
\boldsymbol{Z}_{j(k)} &= \boldsymbol{P}_{j(k)}^{-1} && \text{(7.36g)}
\end{aligned}
$$

for all $\ell \in \{1, \ldots, L+1\}$*,* $j(k) \in \mathbb{M}_0$*, and* $j(k+1) \in \mathbb{M}_0$ *with the unrestricted LMI variables* $\boldsymbol{G}_i \in \mathbb{R}^{n_i \times n_i}$ *and* $\boldsymbol{W}_i \in \mathbb{R}^{m_i \times n_i}$*, the symmetric and positive definite LMI variables* $\boldsymbol{M}_{1i} \in \mathbb{R}^{n_i \times n_i}$*,* $\boldsymbol{M}_{2i} \succ \lambda_i \boldsymbol{M}_{1i}$*,* $\boldsymbol{Z}_{j(k)} \in \mathbb{R}^{2n \times 2n}$*, and* $\beta_{j(k)\ell}^{j(k+1)} \in \mathbb{R}^+$*. The feedback matrices* $\boldsymbol{K}_i$ *in (7.2) and the triggering matrices* $\boldsymbol{S}_i$ *in (7.19) for all* $i \in \mathbb{M}$ *are given by*

$$
\begin{aligned}
\boldsymbol{K}_i &= \boldsymbol{W}_i \boldsymbol{G}_i^{-1} && \text{(7.37a)}\\
\boldsymbol{S}_{1i} &= \boldsymbol{G}_i^{-T} \boldsymbol{M}_{1i} \boldsymbol{G}_i^{-1} && \text{(7.37b)}\\
\boldsymbol{S}_{2i} &= \boldsymbol{G}_i^{-T} \boldsymbol{M}_{2i} \boldsymbol{G}_i^{-1}. && \text{(7.37c)}
\end{aligned}
$$

PROOF. Based on the S-procedure, see Section A.3, the stability condition (7.32) with the restriction that $\boldsymbol{x}(k) \in \mathbb{X}_{j(k)}$ can be formulated with some conservatism as the unrestricted stability condition

$$\Delta V(k) + \boldsymbol{x}^T(k)\tilde{\boldsymbol{S}}_{j(k)}\boldsymbol{x}(k) < -\boldsymbol{x}^T(k)\tilde{\boldsymbol{Q}}_{0j(k)}\boldsymbol{x}(k) \tag{7.38}$$

for all $\boldsymbol{x}(k) \in \mathbb{R}^{2n} \backslash \{\boldsymbol{0}\}$. The proof of the implication of (7.38) by the LMI constraints (7.35b) follows the same line as the proof of Theorem 4.2. For the objective function (7.35a), any of the elements of the set $\mathbb{M}_0$ can be chosen for the switching index $j(0)$. □

Remark 7.8. The LMI optimization problem (7.35) is solved completely offline. Alternatively, we could consider the upper bound (7.33a) as an objective function to the LMI optimization problem (7.35). A reoptimization for the current state $\boldsymbol{x}(k)$, taken as an initial state $\boldsymbol{x}_0$, can then be performed. This is unfortunately not possible, since the *scalar* outputs $\sigma_i(k)$ and not the state vectors of the demanding plants are sent to the scheduler at each time instant t_k. The state measurement $\boldsymbol{x}_{j(k)}(k)$ of the most demanding plant is then transmitted to the corresponding controller at time instant t_k. An online reoptimization at the controller side for the current state measurement $\boldsymbol{x}_{j(k)}(k)$ to improve the resulting control performance is still an open problem.

7.4 Illustrative Example

Consider simultaneous stabilization of three ball-and-beam processes controlled over a single channel communication network. A schematic diagram of the ball-and-beam process is shown in Figure 7.2.

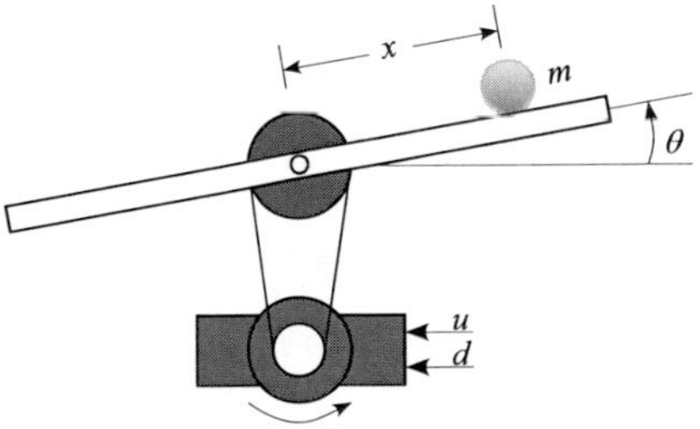

Figure 7.2: Schematic diagram of a ball-and-beam process

Each process consists of a ball with a radius $r_i = 1.5\,\text{cm}$ and a mass $m_i = 0.11\,\text{kg}$, $\forall i \in \mathbb{M} = \{1, 2, 3\}$, rolling on the top of a long beam. The position $x_i(t)$ in meters of the ball is measured with respect to the center of the beam. The beam is mounted at its center on a rotational axis that is driven via a drive belt by an electric DC motor. The DC motor is supplied by an input voltage $u_i(t)$ and subject to a disturbance voltage $d_i(t)$ both in volts.

The continuous-time state equation of each process P_i, assuming that the change of the beam angle $\theta_i(t)$ around the origin is small enough [ÅW97, Example A.1] such that $\sin\theta_i \cong \theta_i$, is given by

$$\underbrace{\begin{pmatrix} \dot{x}_i(t) \\ \ddot{x}_i(t) \end{pmatrix}}_{\dot{\boldsymbol{x}}_{ci}(t)} = \underbrace{\begin{pmatrix} 0 & 1 \\ 0 & 0 \end{pmatrix}}_{\boldsymbol{A}_{ci}} \underbrace{\begin{pmatrix} x_i(t) \\ \dot{x}_i(t) \end{pmatrix}}_{\boldsymbol{x}_{ci}(t)} + \underbrace{\begin{pmatrix} 0 \\ \frac{m_i g k_{\text{u}i}}{(J_i/r_i^2)+m_i} \end{pmatrix}}_{\boldsymbol{b}_{ci}} (u_i(t) + d_i(t)) \tag{7.39}$$

with the gravitational acceleration $g = 9.81\,{}^{\text{m}}\!/\!_{\text{s}^2}$, ball's moment of inertia $J_i = 10^{-5}\,\text{kg.m}^2$, $\forall i \in \mathbb{M}$, and actuator constants $k_{\text{u}1} = 0.3\,{}^{\text{rad}}\!/\!_{\text{V}}$, $k_{\text{u}2} = 0.2\,{}^{\text{rad}}\!/\!_{\text{V}}$, and $k_{\text{u}3} = 0.1\,{}^{\text{rad}}\!/\!_{\text{V}}$ (assuming for simplicity that $\theta_i = k_{\text{u}i}(u_i + d_i)$).

Consider further a single channel communication network with the uncertain time-varying delays

$$\tau_i^{\text{SC}}(k) \in [1\,\text{ms}, 1.5\,\text{ms}] \tag{7.40a}$$

$$\tau_i^{\text{CA}}(k) \in [1\,\text{ms}, 1.5\,\text{ms}] \tag{7.40b}$$

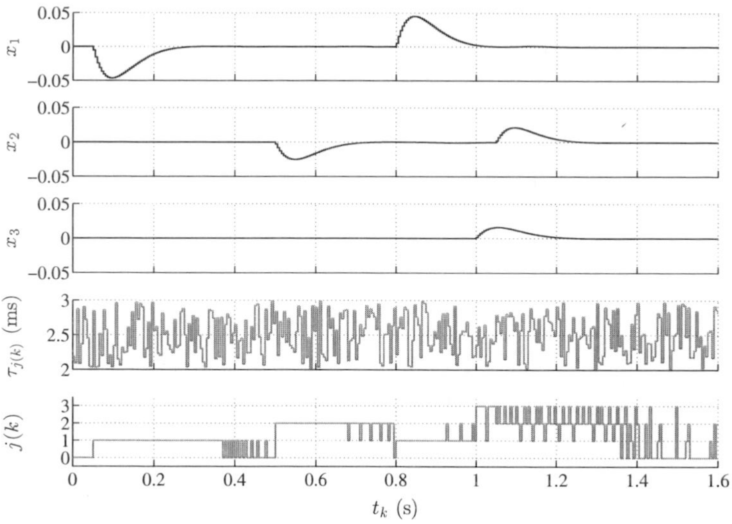

Figure 7.3: Simulation results for tuning parameters $\lambda_i = 0.001$, $\forall i \in \mathbb{M}$, under uniformly distributed impulsive disturbance $d_i(t)$ and random input delay $\tau_i(k)$

and the continuous-time cost function (2.20) with the weighting matrices chosen as

$$\boldsymbol{Q}_{\mathrm{c}i} = \begin{pmatrix} 10^4 & 0 \\ 0 & 10 \end{pmatrix}, \quad R_{\mathrm{c}i} = 10^{-2} \tag{7.41}$$

for all $i \in \mathbb{M}$. The sampling intervals are finally chosen as $h_{j(k)} = h = 5\,\mathrm{ms}$, $\forall j(k) \in \mathbb{M}_0$ following the selection procedure outlined in Section 4.7.

Based on the given parameters of all plants and of the utilized communication network, the parameters of the discrete-time switched polytopic system (7.14) for the approximation order $L = 5$ and of the cost function (7.18) for the nominal input delay $\tau_{0i} = 2.5\,\mathrm{ms}$, $\forall i \in \mathbb{M}$, are first obtained. Problem 7.1 is then solved for the tuning parameters $\lambda_i = 0.001$, $\forall i \in \mathbb{M}$, based on the S-procedure using Theorem 7.1. Simulation results for uniformly distributed impulsive disturbance voltage $d_i(t) \in [\underline{d}_i, \overline{d}_i]$ characterized in Figure 5.2(b) with $\underline{d}_i = -20\,\mathrm{V}$ and $\overline{d}_i = 20\,\mathrm{V}$ and uniformly distributed random input delays $\tau_i(k) \in [2\,\mathrm{ms}, 3\,\mathrm{ms}]$ for all $i \in \mathbb{M}$ are shown in Figure 7.3.

Apparently, the network access is granted to the more demanding plant as seen from the output $j(k)$ of the scheduler. For example, $j(k) = 1$ is decided for $t \in [0.05\,\mathrm{s}, 0.37\,\mathrm{s}]$ since only P_1 is subject to a disturbance impulse. The same behavior results for P_2 within $t \in [0.5\,\mathrm{s}, 0.8\,\mathrm{s}]$ and for P_3 within $t \in [1.0\,\mathrm{s}, 1.05\,\mathrm{s}]$. For $t \in [0.37\,\mathrm{s}, 0.5\,\mathrm{s}]$ and

$t \in [1.35\,\mathrm{s}, 1.6\,\mathrm{s}]$, the idle task T_0 is preferably chosen since all P_i are very close to the equilibrium and hence no necessity for control exists. Thus, an efficient distribution and utilization of the limited communication resource is indeed guaranteed by the proposed event-based switching law (7.22).

For an evaluation of the effect of the chosen tuning parameters λ_i on the resulting control performance and on the network utilization, Problem 7.1 is further solved for different tuning parameters $\lambda_i = \lambda \in \{0.0, 0.05, \ldots, 0.7\}$, $\forall i \in \mathbb{M}$. The resulting cost J in (2.29), averaged over one hundred uniformly distributed random initial ball position $x_{i0} \in [-0.1\,\mathrm{m}, 0.1\,\mathrm{m}]$ and zero disturbance voltage $d_i(t) = 0\ \forall i \in \mathbb{M}$, is depicted in Figure 7.4. We denote by the inter-event time h_{avg} the average of the time interval between two consecutive transmission events.

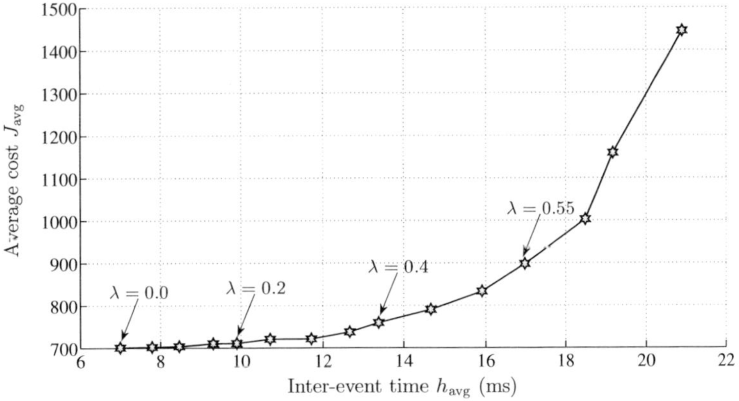

Figure 7.4: Resulting cost for the EBCS strategy, averaged over one hundred uniformly distributed random initial values

As expected, increasing the value of λ reduces the network utilization at the cost of impairing the resulting control performance. The higher the chosen value of λ, the lower the utilization of the communication network and the lower the resulting control performance. Note that even for $\lambda = 0$, the resulting inter-event time is $h_{\text{avg}} = 7\,\mathrm{ms}$ which is larger than the considered sampling interval $h = 5\,\mathrm{ms}$. This is due to the positive definiteness requirement on the weighting matrix $\boldsymbol{S}_{2i}$ in (7.19) that guarantees a certain relative threshold. For $\lambda > 0.7$, the LMI optimization problem (7.35) is not feasible anymore.

Finally, the average of the upper bounds (7.33a) and (7.33b) are depicted in Figure 7.5 to see how far they are from each other and from the resulting average cost J_{avg}. One can easily observe that the upper bound (7.33b) based on trace is much higher than the upper bound (7.33a). Furthermore, the resulting average cost J_{avg} is much less than the upper bound (7.33a), except for short inter-event times. One reason for such huge difference could be that the worst-case scenario for the case of long inter-event times is not triggered in the simulation. Another reason could be because of numerical issues that arise in the LMI optimization problem (7.35) with increasing λ.

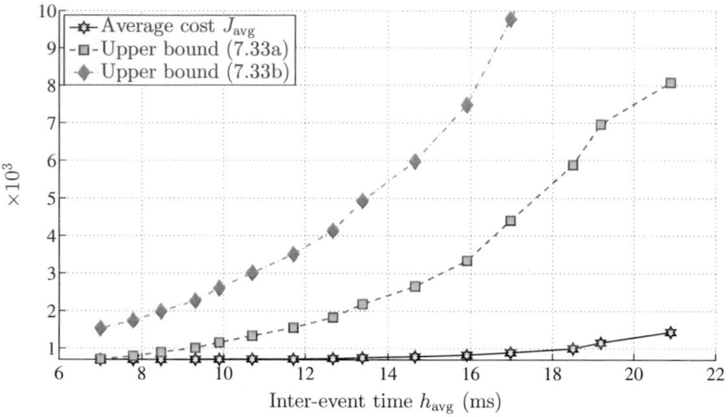

Figure 7.5: Resulting average of the upper bounds (7.33a) and (7.33b)

To sum up, the simulation results have shown the capability of the proposed event-based control and scheduling strategy to systematically balance between the control performance and network utilization, making it an attractive choice for many control applications utilizing general-purpose communication networks.

8 Prediction-Based Control and Scheduling (PBCS)

Within this thesis, the NCS with the restriction that only one plant can get access to the communication network at a time is considered. This is why only one plant is controlled within the sampling interval $t_k \leq t < t_{k+1}$ under all codesign methods proposed in Chapters 4–7. In this chapter, we want to go beyond this classical way of thinking. Strictly speaking, we want to simultaneously control all plants despite the communication restriction. How can we achieve that? The answer is easy, but before answering let's recall the NCS architecture shown in Figure 2.1 or its extended one in Figure 7.1. Although there are M embedded processors, only one of them is operating at a time while the others in the idle mode. Going back to the answer of how all plants can be simultaneously controlled, we consider a modified version of the NCS architecture in Figure 7.1 in which one common embedded processor (rather than M) is used for executing all control tasks T_i, $\forall i \in \mathbb{M}$, based on predicted states. The resulting control signals are then encapsulated in one packet and sent to all plants when necessary. A replica of the predicted states is also generated at the associated event generators for the purpose of evaluating whether the state error (difference between actual state and predicted state) is within a specific range. If not, the actual state measurement of the most demanding plant is sent first to the embedded processor for state-prediction correction purposes. Closed-loop stability and a certain level of control performance can be guaranteed under the proposed prediction-based codesign strategy as shown below.

8.1 NCS Model Extension

A modified version of the NCS architecture shown in Figure 7.1 is depicted in Figure 8.1. Instead of M embedded processors, a single common embedded processor is considered. All control tasks T_i are executed on this common processor based on the outputs of model-based predictors Pr_i, $\forall i \in \mathbb{M}$. The computed control signals are encapsulated in one packet and sent further to all plants when necessary. The necessity of transmitting the computed control signals to the corresponding plants is decided by an associated event generator $\mathrm{EG_u}$. Due to the encapsulation of all control signals, the network-induced delays $\tau_i^{\mathrm{CA}}(k) \in [\underline{\tau}_i^{\mathrm{CA}}, \overline{\tau}_i^{\mathrm{CA}}]$ required for transmitting new control updates to the corresponding plants are assumed to be equal, i.e. $\tau_i^{\mathrm{CA}}(k) = \tau^{\mathrm{CA}}(k)$, $\forall i \in \mathbb{M}$. This technical assumption is mode for simplifying the presentation. The proposed theory

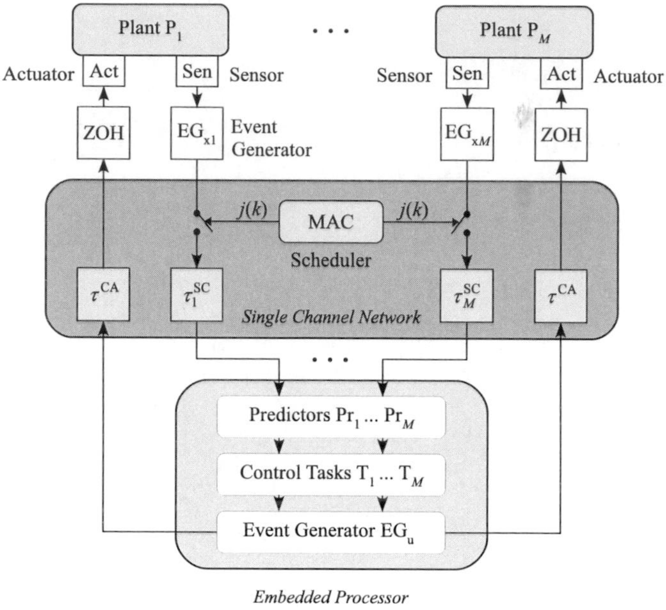

Figure 8.1: Architecture of a prediction-based networked control system

in this chapter can be extended for the non-equal case in a straightforward manner, cf. Remark 8.2. In the following, we derive a mathematical model for the NCS while taking the outlined modifications into account.

Recall the continuous-time state equation (2.1) of plant P_i which is given by

$$\begin{aligned} \dot{\boldsymbol{x}}_{ci}(t) &= \boldsymbol{A}_{ci}\boldsymbol{x}_{ci}(t) + \boldsymbol{B}_{ci}\boldsymbol{u}_i(t) \\ \boldsymbol{x}_{ci}(0) &= \boldsymbol{x}_{ci0}. \end{aligned} \tag{8.1}$$

The state equation (8.1) is discretized over the sampling interval $t_k \le t < t_{k+1}$ using ZOH. Within this interval, the control signal $\boldsymbol{u}_i(t)$ is given by

$$\boldsymbol{u}_i(t) = \begin{cases} \boldsymbol{u}_i(t_{k-1}) & \text{for} \quad t_k \le t < t_k + \tau(k) \\ \boldsymbol{u}_i(t_k) & \text{for} \quad t_k + \tau(k) \le t < t_{k+1} \end{cases} \tag{8.2}$$

where the uncertain time-varying input delay $\tau(k) \in [\underline{\tau}, \overline{\tau}]$ is equal to the network-induced delay $\tau^{\mathrm{CA}}(k)$, i.e. $\tau(k) = \tau^{\mathrm{CA}}(k)$.

Remark 8.1. Since the control tasks $\mathrm{T}_1 \ldots \mathrm{T}_M$ are computed based on predicted states and not based on new state measurements, the network-induced delay $\tau_i^{\mathrm{SC}}(k)$ required for transmitting new state measurements is thereby compensated by the predictor. This is why the control signal in (8.2) experiences only the network-induced delay $\tau^{\mathrm{CA}}(k)$.

Considering (8.2), the solution of the continuous-time state equation (8.1) at sampling instants t_k, $k \in \mathbb{N}$, is given by

$$\boldsymbol{x}_{\mathrm{c}i}(k+1) = \boldsymbol{\Phi}_i(h_{j(k)})\boldsymbol{x}_{\mathrm{c}i}(k) + \boldsymbol{\Gamma}_{1i}(h_{j(k)}, \tau(k))\boldsymbol{u}_i(k-1) + \boldsymbol{\Gamma}_{0i}(h_{j(k)}, \tau(k))\boldsymbol{u}_i(k) \tag{8.3}$$

where $\boldsymbol{\Phi}_i(t)$, $\boldsymbol{\Gamma}_{1i}(t_1, t_2)$, and $\boldsymbol{\Gamma}_{0i}(t_1, t_2)$ are defined according to (2.7). The sampling interval $h_{j(k)} \triangleq t_{k+1} - t_k$ is chosen such that

$$h_{j(k)} \geq \max\{\tau^{\mathrm{SC}}_{j(k)}(k), \tau^{\mathrm{CA}}(k)\} \tag{8.4}$$

is always satisfied. The satisfaction of (8.4) guarantees the transmission of new state measurements and/or new control updates within the chosen sampling interval. Introducing the augmented state vector $\boldsymbol{x}_i(k) = \begin{pmatrix}\boldsymbol{x}^T_{ci}(k) & \boldsymbol{u}^T_i(k-1)\end{pmatrix}^T$, the resulting discrete-time state equation corresponding to (8.1) is thus given by

$$\begin{aligned} \boldsymbol{x}_i(k+1) &= \boldsymbol{A}_{ij(k)}(k)\boldsymbol{x}_i(k) + \boldsymbol{B}_{ij(k)}(k)\boldsymbol{u}_i(k) \\ \boldsymbol{x}_i(0) &= \begin{pmatrix}\boldsymbol{x}^T_{\mathrm{c}i0} & \boldsymbol{0}\end{pmatrix}^T \end{aligned} \tag{8.5}$$

where $\boldsymbol{A}_{ij(k)}(k) \in \mathbb{R}^{(n_i+m_i)\times(n_i+m_i)}$ and $\boldsymbol{B}_{ij(k)}(k) \in \mathbb{R}^{(n_i+m_i)\times m_i}$ are constructed as

$$\boldsymbol{A}_{ij(k)}(k) = \begin{pmatrix}\boldsymbol{\Phi}_i(h_{j(k)}) & \boldsymbol{\Gamma}_i(h_{j(k)}) - \boldsymbol{\Gamma}_i(h_{j(k)} - \tau(k)) \\ \boldsymbol{0} & \boldsymbol{0}\end{pmatrix} \tag{8.6a}$$

$$\boldsymbol{B}_{ij(k)}(k) = \begin{pmatrix}\boldsymbol{\Gamma}_i(h_{j(k)} - \tau(k)) \\ \boldsymbol{I}\end{pmatrix}. \tag{8.6b}$$

Following the overapproximation procedure in Section 2.3 results in a discrete-time plant model with polytopic and additive norm-bounded uncertainty

$$\begin{aligned} \boldsymbol{x}_i(k+1) = &\left(\sum_{\ell=1}^{L+1} \mu_\ell(\rho_{j(k)})\boldsymbol{A}_{ij(k)\ell} + \boldsymbol{D}_{ij(k)}\boldsymbol{E}_{ij(k)}\boldsymbol{F}_{\mathrm{a}i}\right)\boldsymbol{x}_i(k) \\ &+ \left(\sum_{\ell=1}^{L+1} \mu_\ell(\rho_{j(k)})\boldsymbol{B}_{ij(k)\ell} + \boldsymbol{D}_{ij(k)}\boldsymbol{E}_{ij(k)}\boldsymbol{F}_{\mathrm{b}i}\right)\boldsymbol{u}_i(k) \end{aligned} \tag{8.7}$$

where $\rho_{j(k)} = h_{j(k)} - \tau(k)$ and the other parameters are constructed similar to (2.17).

So far, we have formally modeled the dynamics of each plant P_i taking the uncertain time-varying input delay $\tau(k)$ into account. To get a closed-loop model of each plant, we further consider for each control task T_i a prediction-based control law

$$\check{\boldsymbol{u}}_i(k) = \boldsymbol{K}_i\check{\boldsymbol{x}}_i(k) \tag{8.8}$$

where $\check{\boldsymbol{x}}_i(k) \in \mathbb{R}^{n_i+m_i}$ is the predicted state and $\boldsymbol{K}_i \in \mathbb{R}^{m_i \times (n_i+m_i)}$ is the feedback matrix to be determined. The dynamics of the model-based predictor Pr_i is given by

$$\check{\boldsymbol{x}}_i(k+1) = \begin{cases} \boldsymbol{A}^{\mathrm{nom}}_{ij(k)}\boldsymbol{x}_i(k) + \boldsymbol{B}^{\mathrm{nom}}_{ij(k)}\boldsymbol{u}_i(k) & \text{if } j(k) = i \\ \boldsymbol{A}^{\mathrm{nom}}_{ij(k)}\check{\boldsymbol{x}}_i(k) + \boldsymbol{B}^{\mathrm{nom}}_{ij(k)}\boldsymbol{u}_i(k) & \text{otherwise} \end{cases} \tag{8.9}$$

where $\boldsymbol{A}^{\mathrm{nom}}_{ij(k)}$ and $\boldsymbol{B}^{\mathrm{nom}}_{ij(k)}$ are defined in (8.6) after replacing $\tau(k)$ by a nominal input delay τ_0 chosen by the designer. Considering the event generator $\mathrm{EG_u}$, the control signal applied to plant P_i at time instant t_k is then given by

$$\boldsymbol{u}_i(k) = \begin{cases} \check{\boldsymbol{u}}_i(k) & \text{when an event occurs} \\ \boldsymbol{u}_i(k-1) & \text{otherwise.} \end{cases} \tag{8.10}$$

Combining the plant dynamics (8.7) with the predictor dynamics (8.9) while taking the control law (8.8) into consideration yields an augmented closed-loop state equation

$$\begin{aligned} \boldsymbol{\xi}_i(k+1) = {} & \left(\sum_{\ell=1}^{L+1} \mu_\ell(\rho_{j(k)})\tilde{\boldsymbol{A}}_{ij(k)\ell} + \tilde{\boldsymbol{B}}_{ij(k)\ell}\tilde{\boldsymbol{K}}_i + \tilde{\boldsymbol{D}}_{ij(k)}\boldsymbol{E}_{ij(k)}\tilde{\boldsymbol{F}}_i \right) \boldsymbol{\xi}_i(k) \\ & + \left(\sum_{\ell=1}^{L+1} \mu_\ell(\rho_{j(k)})\tilde{\boldsymbol{B}}_{ij(k)\ell} + \tilde{\boldsymbol{D}}_{ij(k)}\boldsymbol{E}_{ij(k)}\boldsymbol{F}_{\mathrm{b}i} \right) \boldsymbol{e}_{\mathrm{u}i}(k) \end{aligned} \tag{8.11}$$

where $\boldsymbol{\xi}_i(k) = \begin{pmatrix} \boldsymbol{x}_i^T(k) & \boldsymbol{e}_{\mathrm{x}i}^T(k) \end{pmatrix}^T$, $\boldsymbol{e}_{\mathrm{x}i}(k) = \boldsymbol{x}_i(k) - \check{\boldsymbol{x}}_i(k)$, $\boldsymbol{e}_{\mathrm{u}i}(k) = \boldsymbol{u}_i(k) - \check{\boldsymbol{u}}_i(k)$, and

$$\tilde{\boldsymbol{A}}_{ij(k)\ell} = \begin{pmatrix} \boldsymbol{A}_{ij(k)\ell} & \boldsymbol{0} \\ \boldsymbol{A}_{ij(k)\ell} - \boldsymbol{A}^{\mathrm{nom}}_{ij(k)} & (1-\delta_{ij(k)})\boldsymbol{A}^{\mathrm{nom}}_{ij(k)} \end{pmatrix} \tag{8.12a}$$

$$\tilde{\boldsymbol{B}}_{ij(k)\ell} = \begin{pmatrix} \boldsymbol{B}_{ij(k)\ell} \\ \boldsymbol{B}_{ij(k)\ell} - \boldsymbol{B}^{\mathrm{nom}}_{ij(k)} \end{pmatrix} \tag{8.12b}$$

$$\tilde{\boldsymbol{D}}_{ij(k)} = \begin{pmatrix} \boldsymbol{D}^T_{ij(k)} & \boldsymbol{D}^T_{ij(k)} \end{pmatrix}^T \tag{8.12c}$$

$$\tilde{\boldsymbol{F}}_i = \begin{pmatrix} \boldsymbol{F}_{\mathrm{a}i} & \boldsymbol{0} \end{pmatrix} + \boldsymbol{F}_{\mathrm{b}i}\tilde{\boldsymbol{K}}_i \tag{8.12d}$$

$$\tilde{\boldsymbol{K}}_i = \begin{pmatrix} \boldsymbol{K}_i & -\boldsymbol{K}_i \end{pmatrix}. \tag{8.12e}$$

The logical variable $\delta_{ij(k)}$ is defined according to (2.5). The NCS model is then given by a block-diagonal discrete-time switched polytopic system

$$\boldsymbol{\xi}(k+1) = \boldsymbol{\mathcal{A}}_{j(k)}(k)\boldsymbol{\xi}(k) + \boldsymbol{\mathcal{B}}_{j(k)}(k)\boldsymbol{e}_{\mathrm{u}}(k) \tag{8.13}$$

where

$$\boldsymbol{\mathcal{A}}_{j(k)}(k) = \sum_{\ell=1}^{L+1} \mu_\ell(\rho_{j(k)})\tilde{\boldsymbol{A}}_{j(k)\ell} + \tilde{\boldsymbol{B}}_{j(k)\ell}\tilde{\boldsymbol{K}} + \tilde{\boldsymbol{D}}_{j(k)}\boldsymbol{E}_{j(k)}\tilde{\boldsymbol{F}} \tag{8.14a}$$

$$\boldsymbol{\mathcal{B}}_{j(k)}(k) = \sum_{\ell=1}^{L+1} \mu_\ell(\rho_{j(k)})\tilde{\boldsymbol{B}}_{j(k)\ell} + \tilde{\boldsymbol{D}}_{j(k)}\boldsymbol{E}_{j(k)}\boldsymbol{F}_{\mathrm{b}}. \tag{8.14b}$$

Remark 8.2. The NCS model (8.13) is derived assuming that the network-induced delays $\tau_i^{\mathrm{CA}}(k)$ at time instant t_k are equal, i.e. $\tau_i^{\mathrm{CA}}(k) = \tau^{\mathrm{CA}}(k)$, $\forall i \in \mathbb{M}$. If this is not the case, the parameters in (8.14) are then modified as

$$\boldsymbol{\mathcal{A}}_{j(k)}(k) = \sum_{\ell_1=1}^{L+1} \mu_{\ell_1}(\rho_{1j(k)}) \cdots \sum_{\ell_M=1}^{L+1} \mu_{\ell_M}(\rho_{Mj(k)}) \tilde{\boldsymbol{A}}_{j(k)\ell_1\ldots\ell_M} + \tilde{\boldsymbol{B}}_{j(k)\ell_1\ldots\ell_M} \tilde{\boldsymbol{K}} + \tilde{\boldsymbol{D}}_{j(k)} \boldsymbol{E}_{j(k)} \tilde{\boldsymbol{F}}$$

$$\boldsymbol{\mathcal{B}}_{j(k)}(k) = \sum_{\ell_1-1}^{L+1} \mu_{\ell_1}(\rho_{1j(k)}) \cdots \sum_{\ell_M=1}^{L+1} \mu_{\ell_M}(\rho_{Mj(k)}) \tilde{\boldsymbol{B}}_{j(k)\ell_1\ldots\ell_M} + \tilde{\boldsymbol{D}}_{j(k)} \boldsymbol{E}_{j(k)} \boldsymbol{F}_{\mathrm{b}}.$$

with the uncertain parameter $\rho_{ij(k)} = h_{j(k)} - \tau_i(k)$ and $\tau_i(k) = \tau_i^{\mathrm{CA}}(k)$ for all $i \in \mathbb{M}$.

For the cost function assigned to each plant P_i, recall further the individual cost function (2.26) with $\hat{\boldsymbol{u}}_i(k) = \boldsymbol{u}_i(k-1)$ and $\tau_{ij(k)}(k)$ replaced by the nominal input delay τ_0, i.e.

$$J_{0i} = \sum_{k=0}^{\infty} \begin{pmatrix} \boldsymbol{x}_{\mathrm{c}i}(k) \\ \boldsymbol{u}_i(k-1) \\ \boldsymbol{u}_i(k) \end{pmatrix}^T \boldsymbol{Q}_{0ij(k)} \begin{pmatrix} \boldsymbol{x}_{\mathrm{c}i}(k) \\ \boldsymbol{u}_i(k-1) \\ \boldsymbol{u}_i(k) \end{pmatrix}. \tag{8.15}$$

The quadratic cost function (8.15) can equivalently be rewritten in terms of the augmented state vector $\boldsymbol{\xi}_i(k)$ and the input error vector $\boldsymbol{e}_{\mathrm{u}i}(k)$ as

$$J_{0i} = \sum_{k=0}^{\infty} \begin{pmatrix} \boldsymbol{\xi}_i(k) \\ \boldsymbol{e}_{\mathrm{u}i}(k) \end{pmatrix}^T \begin{pmatrix} \tilde{\boldsymbol{I}}_i & \boldsymbol{0} \\ \tilde{\boldsymbol{K}}_i & \boldsymbol{I} \end{pmatrix}^T \boldsymbol{Q}_{0ij(k)} \begin{pmatrix} \tilde{\boldsymbol{I}}_i & \boldsymbol{0} \\ \tilde{\boldsymbol{K}}_i & \boldsymbol{I} \end{pmatrix} \begin{pmatrix} \boldsymbol{\xi}_i(k) \\ \boldsymbol{e}_{\mathrm{u}i}(k) \end{pmatrix}. \tag{8.16}$$

where $\tilde{\boldsymbol{I}}_i = \begin{pmatrix} \boldsymbol{I} & \boldsymbol{0} \end{pmatrix}$. The closed-loop cost function assigned to the NCS is thus given by

$$J_0 = \sum_{i=1}^{M} J_{0i} = \sum_{k=0}^{\infty} \begin{pmatrix} \boldsymbol{\xi}(k) \\ \boldsymbol{e}_{\mathrm{u}}(k) \end{pmatrix}^T \underbrace{\begin{pmatrix} \tilde{\boldsymbol{I}} & \boldsymbol{0} \\ \tilde{\boldsymbol{K}} & \boldsymbol{I} \end{pmatrix}^T \boldsymbol{Q}_{0j(k)} \begin{pmatrix} \tilde{\boldsymbol{I}} & \boldsymbol{0} \\ \tilde{\boldsymbol{K}} & \boldsymbol{I} \end{pmatrix}}_{\tilde{\boldsymbol{Q}}_{0j(k)}} \begin{pmatrix} \boldsymbol{\xi}(k) \\ \boldsymbol{e}_{\mathrm{u}}(k) \end{pmatrix} \tag{8.17}$$

with the matrices $\tilde{\boldsymbol{I}}$ and $\tilde{\boldsymbol{K}}$ being block-diagonal and $\boldsymbol{Q}_{0j(k)}$ resulting by construction.

8.2 Problem Formulation

In order to formally describe the prediction-based codesign problem, the architecture of the event generators $\mathrm{EG}_{\mathrm{x}i}$ and EG_{u} is first depicted in Figure 8.2. The structure of each event-triggering law $\sigma_{\mathrm{x}i}(k)$ is given by a quadratic form

$$\sigma_{\mathrm{x}i}(k) \triangleq \begin{pmatrix} \boldsymbol{x}_i(k) \\ \boldsymbol{e}_{\mathrm{x}i}(k) \end{pmatrix}^T \underbrace{\begin{pmatrix} -\boldsymbol{S}_{2i} & \boldsymbol{0} \\ \boldsymbol{0} & \boldsymbol{S}_{1i} \end{pmatrix}}_{\boldsymbol{S}_i} \begin{pmatrix} \boldsymbol{x}_i(k) \\ \boldsymbol{e}_{\mathrm{x}i}(k) \end{pmatrix} > 0 \tag{8.18}$$

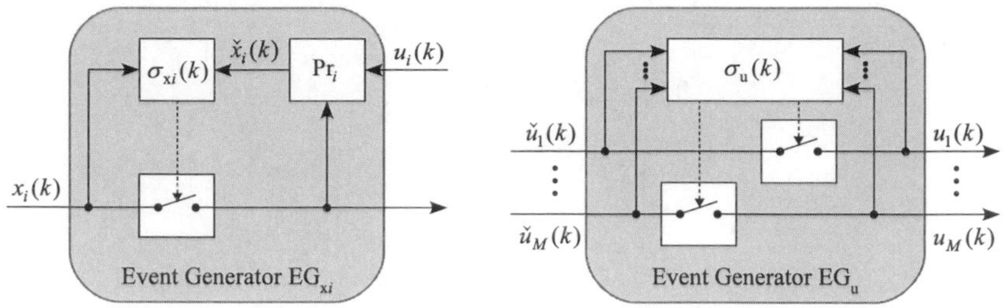

Figure 8.2: Architecture of the event generators $\text{EG}_{\text{x}i}$ and EG_{u}

with the weighting matrices $\boldsymbol{S}_{1i} \succ \boldsymbol{0}$ and $\boldsymbol{S}_{2i} \succ \lambda_{\text{x}i}\boldsymbol{S}_{1i}$, $\forall i \in \mathbb{M}$, to be determined and the tuning parameter $\lambda_{\text{x}i} \geq 0$ for balancing between control performance and network utilization. For further details on the types of event-triggering mechanisms proposed in literature refer to Section 7.2.

Remark 8.3. The event-triggering law (8.18) can equivalently be rewritten as

$$\boldsymbol{e}_{\text{x}i}^T(k)\boldsymbol{S}_{1i}\boldsymbol{e}_{\text{x}i}(k) > \boldsymbol{x}_i^T(k)\boldsymbol{S}_{2i}\boldsymbol{x}_i(k) \tag{8.19}$$

where the squared weighted Euclidean norm of the state error $\boldsymbol{e}_{\text{x}i}(k)$ is compared with the squared weighted Euclidean norm of the state $\boldsymbol{x}_i(k)$ at each time instant t_k. The state-dependent term $\boldsymbol{x}_i^T(k)\boldsymbol{S}_{2i}\boldsymbol{x}_i(k)$ serves as a relative threshold that can be tuned via the design parameter $\lambda_{\text{x}i}$. The higher the chosen value of $\lambda_{\text{x}i}$, the higher the resulting threshold and usually the lower the utilization of the communication network.

A similar quadratic structure is further considered for the event-triggering law $\sigma_{\text{u}}(k)$ and is given by

$$\big(\boldsymbol{u}(k-1) - \check{\boldsymbol{u}}(k)\big)^T \boldsymbol{R}_1 \big(\boldsymbol{u}(k-1) - \check{\boldsymbol{u}}(k)\big) > \check{\boldsymbol{u}}^T(k)\boldsymbol{R}_2\check{\boldsymbol{u}}(k) \tag{8.20}$$

with the weighting matrices $\boldsymbol{R}_1 \succ \boldsymbol{0}$ and $\boldsymbol{R}_2 \succ \lambda_{\text{u}}\boldsymbol{R}_1$ to be determined and the tuning parameter $\lambda_{\text{u}} \geq 0$ for balancing between control performance and network utilization.

Remark 8.4. The control signal $\check{\boldsymbol{u}}_i(k)$ in (8.8) is computed at each time instant t_k based on the predicted state signal $\check{\boldsymbol{x}}_i(k)$. Two event generators are thus required in each control loop, one for the decision whether to send the current state measurement $\boldsymbol{x}_i(k)$ to the corresponding predictor Pr_i while the other one for the decision whether to send $\check{\boldsymbol{u}}_i(k)$ to the corresponding plant.

Motivated by the proposed event-triggering law (8.18), consider finally the scheduler implementing an event-based switching law

$$j(k) = \begin{cases} 0 & \text{if } \sigma_{\text{x}i}(k) \leq 0,\ \forall i \in \mathbb{M} \\ \arg\max\limits_{i \in \mathbb{M}} \sigma_{\text{x}i}(k) & \text{otherwise.} \end{cases} \tag{8.21}$$

The prediction-based codesign problem of the control law (8.8) and the switching law (8.21) can now be formulated as

Problem 8.1 *For the closed-loop switched polytopic system* (8.13) *find the feedback matrices* $\boldsymbol{K}_i$ *in* (8.8) *and the triggering matrices* $\boldsymbol{S}_i$ *in* (8.18) *and* $\boldsymbol{R}_1$, $\boldsymbol{R}_2$ *in* (8.20) *for all* $i \in \mathbb{M}$ *such that the closed-loop cost function* (8.17) *is robustly minimized for all* $\tau(k) \in [\underline{\tau}, \overline{\tau}]$, *i.e.*

$$\min_{\substack{\boldsymbol{R}_1,\, \boldsymbol{R}_2 \\ \boldsymbol{K}_i,\, \boldsymbol{S}_i\ \forall i \in \mathbb{M}}} J_0^{\max} \quad \text{subject to (8.13), (8.20), and (8.21)} \tag{8.22}$$

where

$$J_0^{\max} \triangleq \max_{\tau(0),\ldots,\tau(\infty)} J_0. \tag{8.23}$$

Remark 8.5. According to Remark 4.1, Problem 8.1 is computationally intractable and thus an upper bound on the objective function (8.23) is derived in the following. The resulting upper bound is then considered as a new objective function for a tractable optimization problem.

8.3 Solution based on the S-Procedure

Consider a switched Lyapunov function

$$V(k) = \boldsymbol{\xi}^T(k) \boldsymbol{P}_{j(k)} \boldsymbol{\xi}(k) \tag{8.24}$$

where the Lyapunov matrices $\boldsymbol{P}_{j(k)} \in \mathbb{R}^{(2n+2m)\times(2n+2m)}$, $\forall j(k) \in \mathbb{M}_0$, are symmetric and positive definite. The set $\mathbb{M}_0$ is defined according to (7.34). The difference of the Lyapunov function, i.e. $\Delta V(k) \triangleq V(k+1) - V(k)$, along trajectories of the closed-loop switched polytopic system (8.13) is given by

$$\Delta V(k) = \begin{pmatrix} \boldsymbol{\xi}(k) \\ \boldsymbol{e}_{\mathrm{u}}(k) \end{pmatrix}^T \left[\begin{pmatrix} \boldsymbol{\mathcal{A}}_{j(k)}^T(k) \\ \boldsymbol{\mathcal{B}}_{j(k)}^T(k) \end{pmatrix} \boldsymbol{P}_{j(k+1)} \begin{pmatrix} \boldsymbol{\mathcal{A}}_{j(k)}(k) & \boldsymbol{\mathcal{B}}_{j(k)}(k) \end{pmatrix} - \begin{pmatrix} \boldsymbol{P}_{j(k)} & \boldsymbol{0} \\ \boldsymbol{0} & \boldsymbol{0} \end{pmatrix} \right] \begin{pmatrix} \boldsymbol{\xi}(k) \\ \boldsymbol{e}_{\mathrm{u}}(k) \end{pmatrix}. \tag{8.25}$$

Lemma 8.1 *Based on the event-triggering law* (8.20), *the input error vector* $\mathbf{e}_{\mathrm{u}}(k)$ *at each time instant* t_k *is bounded, i.e.* $\mathbf{e}_{\mathrm{u}}(k) \in \mathbb{U}_k \subset \mathbb{R}^m$, *with the set* $\mathbb{U}_k$ *given by*

$$\mathbb{U}_k \triangleq \{\boldsymbol{e}_{\mathrm{u}}(k) \in \mathbb{R}^m \mid \boldsymbol{e}_{\mathrm{u}}^T(k) \boldsymbol{R}_1 \boldsymbol{e}_{\mathrm{u}}(k) \leq \boldsymbol{\xi}^T(k) \tilde{\boldsymbol{K}}^T \boldsymbol{R}_2 \tilde{\boldsymbol{K}} \boldsymbol{\xi}(k)\}. \tag{8.26}$$

The matrices $\boldsymbol{R}_1$ *and* $\boldsymbol{R}_2$ *are defined according to* (8.20).

PROOF. From (8.10) we obtain

$$\boldsymbol{u}(k) = \begin{cases} \breve{\boldsymbol{u}}(k) & \text{when (8.20) is satisfied} \\ \boldsymbol{u}(k-1) & \text{otherwise.} \end{cases} \tag{8.27}$$

In the case that (8.20) is satisfied, we have $\boldsymbol{u}(k) = \check{\boldsymbol{u}}(k)$ and hence it holds that

$$\left(\boldsymbol{u}(k) - \check{\boldsymbol{u}}(k)\right)^T \boldsymbol{R}_1 \underbrace{\left(\boldsymbol{u}(k) - \check{\boldsymbol{u}}(k)\right)}_{=\boldsymbol{0}} \leq \check{\boldsymbol{u}}^T(k)\boldsymbol{R}_2\check{\boldsymbol{u}}(k). \tag{8.28}$$

In the case that (8.20) is not satisfied, we have $\boldsymbol{u}(k) = \boldsymbol{u}(k-1)$ and hence it holds that

$$\left(\boldsymbol{u}(k) - \check{\boldsymbol{u}}(k)\right)^T \boldsymbol{R}_1 \left(\boldsymbol{u}(k) - \check{\boldsymbol{u}}(k)\right) \leq \check{\boldsymbol{u}}^T(k)\boldsymbol{R}_2\check{\boldsymbol{u}}(k). \tag{8.29}$$

Based on (8.28) and (8.29), we arrive at the conclusion that

$$\begin{aligned}\boldsymbol{e}_{\mathrm{u}}^T(k)\boldsymbol{R}_1\boldsymbol{e}_{\mathrm{u}}(k) &\leq \check{\boldsymbol{u}}^T(k)\boldsymbol{R}_2\check{\boldsymbol{u}}(k)\\ &= \check{\boldsymbol{x}}^T(k)\boldsymbol{K}^T\boldsymbol{R}_2\boldsymbol{K}\check{\boldsymbol{x}}(k)\\ &= \boldsymbol{\xi}^T(k)\tilde{\boldsymbol{K}}^T\boldsymbol{R}_2\tilde{\boldsymbol{K}}\boldsymbol{\xi}(k)\end{aligned} \tag{8.30}$$

for all $k \in \mathbb{N}_0$ with $\boldsymbol{K} = \mathrm{diag}(\boldsymbol{K}_1, \ldots, \boldsymbol{K}_M)$. This completes the proof. □

Lemma 8.2 *Based on the event-based switching law* (8.21)*, the state space* $\mathbb{R}^{2n+2m}$ *is partitioned into regions each expressed by a quadratic form*

$$\mathbb{X}_{j(k)} \triangleq \{\boldsymbol{\xi}(k) \in \mathbb{R}^{2n+2m} \mid \boldsymbol{\xi}^T(k)\tilde{\boldsymbol{S}}_{j(k)}\boldsymbol{\xi}(k) \geq 0\} \tag{8.31}$$

where

$$\tilde{\boldsymbol{S}}_{j(k)} = \begin{cases}\mathrm{diag}\left(-\boldsymbol{S}_1, \ldots, -\boldsymbol{S}_M\right) & \text{for } j(k) = 0\\ \mathrm{diag}\left(-\boldsymbol{S}_1, \ldots, (M-1)\boldsymbol{S}_{j(k)}, \ldots, -\boldsymbol{S}_M\right) & \text{otherwise.}\end{cases} \tag{8.32}$$

Proof. Follows the same line as the proof of Lemma 7.1. □

Lemma 8.3 *If the difference of the Lyapunov function* (8.25) *satisfies*

$$\Delta V(k) < -\begin{pmatrix}\boldsymbol{\xi}(k)\\ \boldsymbol{e}_u(k)\end{pmatrix}^T \tilde{\boldsymbol{Q}}_{0j(k)} \begin{pmatrix}\boldsymbol{\xi}(k)\\ \boldsymbol{e}_u(k)\end{pmatrix} \qquad \forall \boldsymbol{\xi}(k) \in \mathbb{X}_{j(k)} \quad \forall \boldsymbol{e}_u(k) \in \mathbb{U}_k \tag{8.33}$$

with $\tilde{\boldsymbol{Q}}_{0j(k)}$ *defined in* (8.17)*, then the objective function* (8.23) *is upper bounded by*

$$J_0^{\max} < \boldsymbol{\xi}^T(0)\boldsymbol{P}_{j(0)}\boldsymbol{\xi}(0) \tag{8.34a}$$

$$< \mathrm{tr}\left(\boldsymbol{P}_{j(0)}\right)\boldsymbol{\xi}^T(0)\boldsymbol{\xi}(0). \tag{8.34b}$$

Proof. The proof follows the same line as the proof of Theorem 4.1. The additional restrictions, namely $\boldsymbol{\xi}(k) \in \mathbb{X}_{j(k)}$ and $\boldsymbol{e}_{\mathrm{u}}(k) \in \mathbb{U}_k$, are induced due to the event-based switching law (8.21) and the event-triggering law (8.20), respectively. □

The LMI optimization problem corresponding to Problem 8.1 can now be formulated as

Theorem 8.1 *The solution to Problem 8.1 with the objective function* (8.23) *replaced by the upper bound* (8.34b) *results from the LMI optimization problem*

$$\min_{\boldsymbol{G},\boldsymbol{W},\boldsymbol{Y},\tilde{\boldsymbol{R}}_1,\tilde{\boldsymbol{R}}_2,\boldsymbol{Z}_{j(k)},\boldsymbol{M}_{j(k)},\beta_{j(k)\ell}^{j(k+1)}} \operatorname{tr}\left(\boldsymbol{P}_{j(0)}\right) \quad \text{subject to} \tag{8.35a}$$

$$\begin{pmatrix} \boldsymbol{G}^T+\boldsymbol{G}-\boldsymbol{Z}_{j(k)}-\boldsymbol{M}_{j(k)} & * & * & * & * & * \\ \boldsymbol{0} & \boldsymbol{Y}^T+\boldsymbol{Y}-\tilde{\boldsymbol{R}}_1 & * & * & * & * \\ \tilde{\boldsymbol{A}}_{j(k)\ell}\boldsymbol{G}+\tilde{\boldsymbol{B}}_{j(k)\ell}\boldsymbol{W} & \tilde{\boldsymbol{B}}_{j(k)\ell}\boldsymbol{Y} & \boldsymbol{X}_{j(k)\ell}^{j(k+1)} & * & * & * \\ \boldsymbol{W} & \boldsymbol{0} & \boldsymbol{0} & \tilde{\boldsymbol{R}}_2 & * & * \\ \boldsymbol{Q}_{j(k)}^{1/2}\begin{pmatrix}\tilde{\boldsymbol{I}}\boldsymbol{G}\\ \boldsymbol{W}\end{pmatrix} & \boldsymbol{Q}_{j(k)}^{1/2}\begin{pmatrix}\boldsymbol{0}\\ \boldsymbol{Y}\end{pmatrix} & \boldsymbol{0} & \boldsymbol{0} & \boldsymbol{I} & * \\ \tilde{\boldsymbol{F}}_{\mathrm{a}}\boldsymbol{G}+\boldsymbol{F}_{\mathrm{b}}\boldsymbol{W} & \boldsymbol{F}_{\mathrm{b}}\boldsymbol{Y} & \boldsymbol{0} & \boldsymbol{0} & \boldsymbol{0} & \beta_{j(k)\ell}^{j(k+1)}\boldsymbol{I} \end{pmatrix} \succ \boldsymbol{0} \tag{8.35b}$$

where

$$\boldsymbol{G} = \operatorname{diag}\left(\boldsymbol{G}_1, \boldsymbol{G}_1, \ldots, \boldsymbol{G}_M, \boldsymbol{G}_M\right) \tag{8.36a}$$

$$\boldsymbol{W} = \operatorname{diag}\left(\left(\boldsymbol{W}_1 \quad -\boldsymbol{W}_1\right), \ldots, \left(\boldsymbol{W}_M \quad -\boldsymbol{W}_M\right)\right) \tag{8.36b}$$

$$\tilde{\boldsymbol{F}}_{\mathrm{a}} = \operatorname{diag}\left(\left(\boldsymbol{F}_{\mathrm{a}1} \quad \boldsymbol{0}\right), \ldots, \left(\boldsymbol{F}_{\mathrm{a}M} \quad \boldsymbol{0}\right)\right) \tag{8.36c}$$

$$\boldsymbol{Z}_{j(k)} = \boldsymbol{P}_{j(k)}^{-1}, \quad \tilde{\boldsymbol{R}}_1 = \boldsymbol{R}_1^{-1}, \quad \tilde{\boldsymbol{R}}_2 = \boldsymbol{R}_2^{-1} \tag{8.36d}$$

$$\tilde{\boldsymbol{M}}_{j(k)} = \begin{cases} \operatorname{diag}\left(-\boldsymbol{M}_1, \ldots, -\boldsymbol{M}_M\right) & \text{for } j(k) = 0 \\ \operatorname{diag}\left(-\boldsymbol{M}_1, \ldots, (M-1)\boldsymbol{M}_{j(k)}, \ldots, -\boldsymbol{M}_M\right) & \text{otherwise} \end{cases} \tag{8.36e}$$

$$\boldsymbol{M}_i = \operatorname{diag}\left(-\boldsymbol{M}_{2i}, \boldsymbol{M}_{1i}\right) \quad \forall i \in \mathbb{M} \tag{8.36f}$$

$$\boldsymbol{X}_{j(k)\ell}^{j(k+1)} = \boldsymbol{Z}_{j(k+1)} - \beta_{j(k)\ell}^{j(k+1)}\tilde{\boldsymbol{D}}_{j(k)}\tilde{\boldsymbol{D}}_{j(k)}^T \tag{8.36g}$$

for all $\ell \in \{1, \ldots, L+1\}$, $j(k) \in \mathbb{M}_0$, *and* $j(k+1) \in \mathbb{M}_0$ *with the unrestricted LMI variables* $\boldsymbol{G}_i \in \mathbb{R}^{(n_i+m_i)\times(n_i+m_i)}$, $\boldsymbol{W}_i \in \mathbb{R}^{m_i\times(n_i+m_i)}$, *and* $\boldsymbol{Y} \in \mathbb{R}^{m\times m}$, *the symmetric and positive definite LMI variables* $\boldsymbol{M}_{1i} \in \mathbb{R}^{(n_i+m_i)\times(n_i+m_i)}$, $\boldsymbol{M}_{2i} \succ \lambda_{\mathrm{x}i}\boldsymbol{M}_{1i}$, $\tilde{\boldsymbol{R}}_2 \in \mathbb{R}^{m\times m}$, $\tilde{\boldsymbol{R}}_1 \succ \lambda_{\mathrm{u}}\tilde{\boldsymbol{R}}_2$, $\boldsymbol{Z}_{j(k)} \in \mathbb{R}^{(2n+2m)\times(2n+2m)}$, *and* $\beta_{j(k)\ell}^{j(k+1)} \in \mathbb{R}^+$. *The feedback matrices* $\boldsymbol{K}_i$ *in* (8.8) *and the triggering matrices* $\boldsymbol{S}_i$ *in* (8.18) *and* $\boldsymbol{R}_1$, $\boldsymbol{R}_2$ *in* (8.20) *for all* $i \in \mathbb{M}$ *are given by*

$$\boldsymbol{K}_i = \boldsymbol{W}_i\boldsymbol{G}_i^{-1} \tag{8.37a}$$

$$\boldsymbol{R}_1 = \tilde{\boldsymbol{R}}_1^{-1}, \qquad \boldsymbol{R}_2 = \tilde{\boldsymbol{R}}_2^{-1} \tag{8.37b}$$

$$\boldsymbol{S}_{1i} = \boldsymbol{G}_i^{-T}\boldsymbol{M}_{1i}\boldsymbol{G}_i^{-1}, \quad \boldsymbol{S}_{2i} = \boldsymbol{G}_i^{-T}\boldsymbol{M}_{2i}\boldsymbol{G}_i^{-1}. \tag{8.37c}$$

PROOF. Based on the S-procedure, see Section A.3, the stability condition (8.33) with the restrictions that $\boldsymbol{\xi}(k) \in \mathbb{X}_{j(k)}$ and $\boldsymbol{e}_{\mathrm{u}}(k) \in \mathbb{U}_k$ can be formulated with some conservatism as the unrestricted stability condition

$$\Delta V(k) + \boldsymbol{\xi}^T(k)\big(\tilde{\boldsymbol{S}}_{j(k)} + \tilde{\boldsymbol{K}}^T \boldsymbol{R}_2 \tilde{\boldsymbol{K}}\big)\boldsymbol{\xi}(k) - \boldsymbol{e}_{\mathrm{u}}^T(k)\boldsymbol{R}_1 \boldsymbol{e}_{\mathrm{u}}(k) < -\begin{pmatrix}\boldsymbol{\xi}(k)\\ \boldsymbol{e}_{\mathrm{u}}(k)\end{pmatrix}^T \tilde{\boldsymbol{Q}}_{0j(k)} \begin{pmatrix}\boldsymbol{\xi}(k)\\ \boldsymbol{e}_{\mathrm{u}}(k)\end{pmatrix} \tag{8.38}$$

for all $\boldsymbol{\xi}(k) \neq \mathbf{0}$ and $\boldsymbol{e}_{\mathrm{u}}(k) \neq \mathbf{0}$. The proof of the implication of (8.38) by the LMI constraints (8.35b) follows the same line as for Theorem 4.2. For the objective function (7.35a), any of the elements of the set $\mathbb{M}_0$ can be chosen for the switching index $j(0)$. □

Remark 8.6. The LMI optimization problem (8.35) is solved completely offline. An online reoptimization for the current state measurement $\boldsymbol{x}_{j(k)}(k)$, following the setup outlined in Remark 7.8, to improve the control performance is still an open problem.

8.4 Illustrative Example

Consider three line-following mobile robots with the schematic diagram shown in Figure 8.3. Line- or path-following mobile robots can be used in many applications like warehouses and airports. Each robot follows a visual line painted on or embedded in the floor by adjusting its orientation $\theta_i(t)$ in radian and offset $y_i(t)$ in meters, $\forall i \in \mathbb{M} = \{1, 2, 3\}$, with respect to the path using a servomotor mounted on the steering wheels assembly.

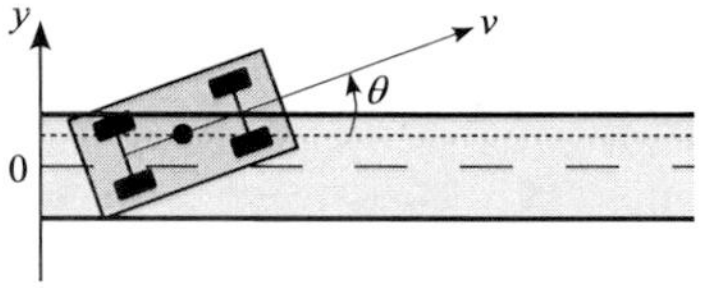

Figure 8.3: Schematic diagram of a line-following mobile robot

The continuous-time state equation of each mobile robot P_i, assuming that the change of the orientation angle $\theta_i(t)$ around the origin is small enough such that $\sin\theta_i(t) \cong \theta_i(t)$, is given by

$$\underbrace{\begin{pmatrix}\dot{y}_i(t)\\ \dot{\theta}_i(t)\end{pmatrix}}_{\dot{\boldsymbol{x}}_{ci}(t)} = \underbrace{\begin{pmatrix}0 & v_i\\ 0 & 0\end{pmatrix}}_{\boldsymbol{A}_{ci}} \underbrace{\begin{pmatrix}y_i(t)\\ \theta_i(t)\end{pmatrix}}_{\boldsymbol{x}_{ci}(t)} + \underbrace{\begin{pmatrix}0\\ 1\end{pmatrix}}_{\boldsymbol{b}_{ci}} \underbrace{w_i(t)}_{u_i(t)} + \begin{pmatrix}0\\ 1\end{pmatrix} d_i(t) \tag{8.39}$$

where $w_i(t)$ is the angular velocity of the servomotor in radian per second and $v_{1,2,3} = 2, 5, 10\,\mathrm{m/s}$ is the linear velocity of the robot. The disturbance input $d_i(t)$ can be interpreted as a disturbance angular velocity in radian per second due to an external force acting on the robot.

Consider further a single channel communication network with the uncertain time-varying delays

$$\tau_i^{\mathrm{SC}}(k) \in [1\,\mathrm{ms}, 4\,\mathrm{ms}] \tag{8.40a}$$

$$\tau^{\mathrm{CA}}(k) \in [1\,\mathrm{ms}, 4\,\mathrm{ms}] \tag{8.40b}$$

and the continuous-time cost function (2.20) with the weighting matrices chosen as

$$\boldsymbol{Q}_{\mathrm{c}i} = \begin{pmatrix} 10^4 & 0 \\ 0 & 0.1 \end{pmatrix}, \quad R_{\mathrm{c}i} = 10^{-2} \tag{8.41}$$

for all $i \in \mathbb{M}$. The sampling intervals are finally chosen as $h_{j(k)} = h = 5\,\mathrm{ms}$, $\forall j(k) \in \mathbb{M}_0$, following the selection procedure given in Section 4.7.

Based on the given parameters of all plants and of the utilized communication network, the parameters of the discrete-time switched polytopic system (8.13) and of the cost function (8.17) for the approximation order $L = 5$ and the nominal input delay $\tau_0 = 2.5\,\mathrm{ms}$ are first obtained. Problem 8.1 is then solved for the tuning parameters $\lambda_{\mathrm{x}i} = \lambda_{\mathrm{u}} = 0.001$, $\forall i \in \mathbb{M}$, based on the S-procedure using Theorem 8.1. Simulation results for uniformly distributed impulsive disturbance inputs $d_i(t) \in [\underline{d}_i, \overline{d}_i]$ characterized in Figure 5.2(b) with $\underline{d}_i = -100\,\mathrm{rad/s}$ and $\overline{d}_i = 100\,\mathrm{rad/s}$ and uniformly distributed random input delays $\tau(k) \in [1\,\mathrm{ms}, 4\,\mathrm{ms}]$ are shown in Figure 8.4. We denote by $\mathrm{EG}_u = 1$ the status where an event is triggered and the control signals of all plants are updated.

Evidently, the proposed event-based switching law (8.21) directly reacts to any disturbance impulse acting on any plant P_i by allowing its current state measurement to be sent to the corresponding predictor Pr_i. For instance, the output of the scheduler is $j(k) = 1$ once the plant P_1 is subject to a disturbance impulse at time instant $t = 0.05\,\mathrm{s}$. The same behavior results for P_3 at time instant $t = 0.25\,\mathrm{s}$ and for P_2 at time instant $t = 0.5\,\mathrm{s}$. The current state measurements are also sent to the corresponding predictors Pr_i even if no disturbances are acting on the plants. This can for instance be seen at time instants $t = 0.4\,\mathrm{s}$ for plant P_1, $t = 0.82\,\mathrm{s}$ for plant P_2, and $t = 0.9\,\mathrm{s}$ for plant P_3. This is in fact due to the consideration of the nominal input delay τ_0 in the predictor dynamical model which leads to a model mismatch compared to the actual plant behavior. Occasional correction of the predicted state value is thus to be expected. The resulting correction rate, i.e. number of state-correction times ($j(k) \neq 0$) to number of simulation iterations, within the simulation time interval $[0\,\mathrm{s}, 2\,\mathrm{sec}]$ is equal to 15%.

For the event generator EG_{u} at the controller side, the same behavior as for the event-based switching law (8.21) can be observed with the difference that all control signals (and not just one of them) are updated in case of an event is triggered. For example, the event generator EG_{u} is activated for $t \in [0.05\,\mathrm{s}, 0.17\,\mathrm{s}]$ since plant P_1 is subject to disturbance. For $t \in [1.55\,\mathrm{s}, 1.75\,\mathrm{s}]$, however, there is almost no activation of the event generator EG_{u} because all plants are very close to the equilibrium and hence no necessity for control update exists. The resulting update rate, i.e. number of control-update times to number of simulation iterations, within the simulation time interval $[0\,\mathrm{s}, 2\,\mathrm{sec}]$ is equal

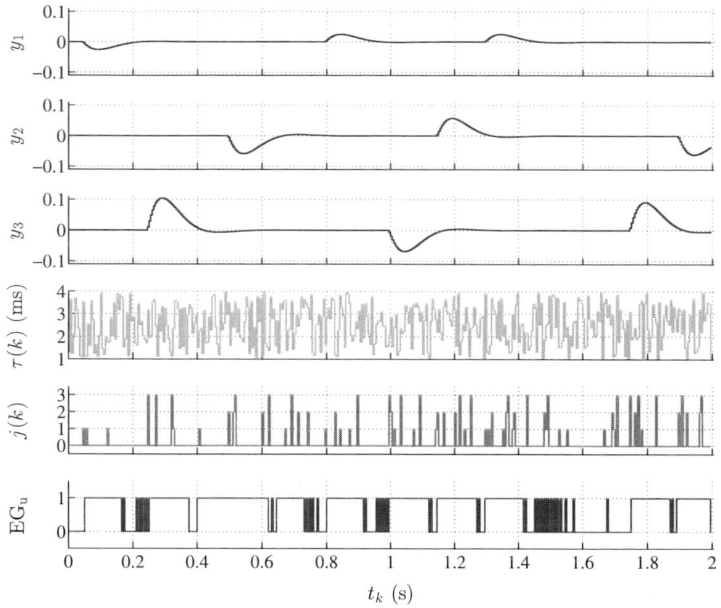

Figure 8.4: Simulation results under uniformly distributed impulsive disturbances and random input delay $\tau(k)$ with the tuning parameters $\lambda_{xi} = \lambda_u = 0.001$, $\forall i$

to 70%. Overall, an efficient distribution and utilization of the limited communication resource is indeed guaranteed by the proposed event generators EG_{xi} and EG_u together with the event-based switching law (8.21).

For an evaluation of the effect of the chosen tuning parameters λ_{xi} and λ_u on the resulting control performance and on the network utilization, Problem 8.1 is further solved for $\lambda_{xi} = \lambda_x \in \{0.001, 0.01, 0.1\}$ and $\lambda_u \in \{0.0, 0.05, \ldots, \lambda_u^{\max}\}$ where $\lambda_u^{\max}$ is the maximum value of λ_u that can be chosen for a given λ_x while guaranteeing the feasibility of the LMI optimization problem (8.35). Surely, $\lambda_u^{\max}$ might be different for different values of λ_x. The resulting cost J in (2.29), averaged over one hundred normally distributed random initial state values $\boldsymbol{x}_{ci0} = \begin{pmatrix} y_{i0} & \theta_{i0} \end{pmatrix}^T$ with zero expected value and unit covariance matrix and zero disturbance inputs $d_i(t) = 0$ for all $i \in \mathbb{M}$, is depicted in Figure 8.5. The inter-event time for a given initial value $\boldsymbol{x}_{ci0}$ is computed as

$$h_{\text{avg}}(\boldsymbol{x}_{ci0}) = \frac{\text{Sampling interval } h = 5\,\text{ms}}{\text{Update rate}}. \tag{8.42}$$

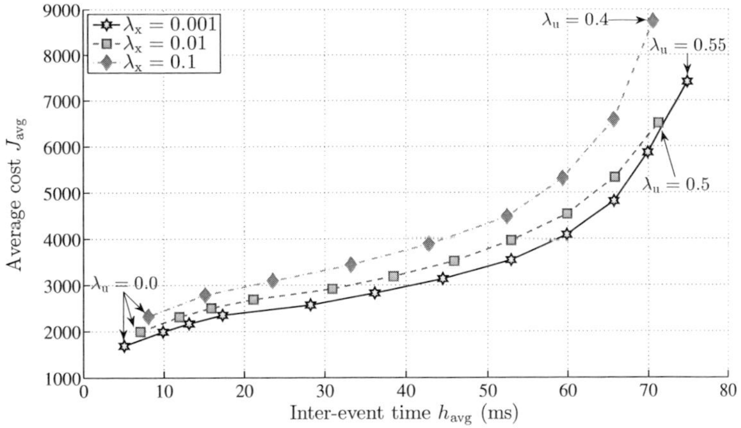

Figure 8.5: Resulting cost for the PBCS strategy, averaged over one hundred normally distributed random initial values

Note that the correction rate is excluded from the calculation of $h_{avg}(\boldsymbol{x}_{ci0})$, focusing only on the controller-actuator communication path.

As expected, increasing the value of λ_u for a given λ_x reduces the network utilization at the cost of degrading the resulting control performance. The higher the chosen value of λ_u, the lower the utilization of the communication network and the lower the control performance. An interesting point to be observed is that the same conclusion can be made for the tuning parameter λ_x under a given λ_u. This is due to the fact that the increase of the value of λ_x reduces the state-correction rate. Consequently, the control signals will be updated for a longer time according to predicted states that might be far away from the actual system state. Moreover, the resulting update rate will be reduced as well since no frequent feedback about the actual system state is obtained. Thus, the utilization of the network is reduced at the cost of more performance degradation.

To sum up, the simulation results have shown the capability of the proposed prediction-based control and scheduling strategy to systematically balance between control performance and network utilization while using a few number of embedded processors. In contrast to the proposed EBCS strategy in Chapter 7, model-based event generators have been proposed for generating state estimates instead of holding last transmitted measurements. Furthermore, the control signals of all plants (rather than only one of them) have been simultaneously updated as a result of the available information on system dynamics at the controller side.

9 Evaluation and Implementation

The control and scheduling codesign strategies proposed in Chapters 4–8 are first evaluated for a simulation study in this chapter. Different simulation scenarios are performed with which the properties of each codesign method as well as the effects of the associated design parameters on the resulting control performance are investigated in more detail. The codesign strategies are next implemented on a benchmark system. Some details on the requirements and the challenges of each codesign strategy from the implementation perspective are also given. The simulation/experimental results under each codesign strategy are finally discussed from which a general conclusion is drawn.

9.1 Evaluation

9.1.1 Simulation Setup

For evaluation purposes, the setup where three actuated pendulums are simultaneously stabilized with the objective of maintaining them in the upright equilibrium $\theta_i = 0\,\text{rad}$, $\forall i \in \mathbb{M} = \{1, 2, 3\}$, is considered in the following. A schematic diagram of the actuated pendulum is shown in Figure 9.1.

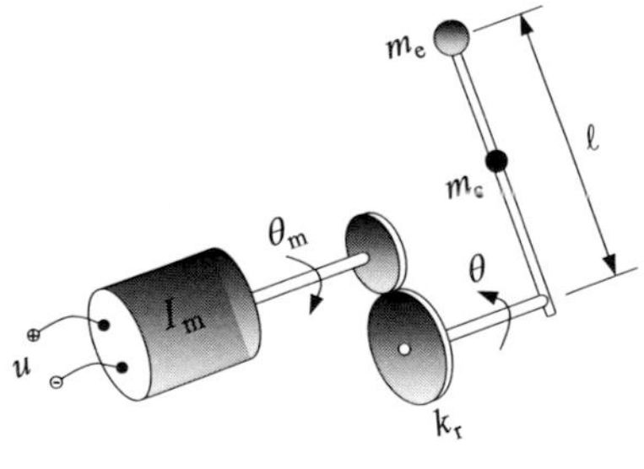

Figure 9.1: Schematic diagram of an actuated pendulum

Actuated pendulums are also known as one-link robot manipulators, see e.g. [Ż03, Example 5.13] where a detailed derivation of the state equation (9.2) is given. The parameters of each pendulum, namely its length ℓ_i, end-point mass m_{ei}, and center-point mass m_{ci}, are given by

$$\ell_1 = 0.25\,\mathrm{m}, \quad m_{e1} = 0.0182\,\mathrm{kg}, \quad m_{c1} = 0.0648\,\mathrm{kg} \tag{9.1a}$$
$$\ell_2 = 0.40\,\mathrm{m}, \quad m_{e2} = 0.0077\,\mathrm{kg}, \quad m_{c2} = 0.0982\,\mathrm{kg} \tag{9.1b}$$
$$\ell_3 = 0.50\,\mathrm{m}, \quad m_{e3} = 0.0318\,\mathrm{kg}, \quad m_{c3} = 0.1207\,\mathrm{kg}. \tag{9.1c}$$

The motion of the pendulum is driven via a gearbox by an armature-controlled DC motor. The DC motors together with the gearboxes of all actuated pendulums are assumed to be identical with the corresponding parameters summarized in Table 9.1.

Parameter	Description	Value	Unit
k_r	Gear ratio	28	–
k_v	Speed constant	0.0258	Vs/rad
k_t	Torque constant	0.0258	Nm/A
R_a	Rotor resistance	3.26	Ω
L_a	Rotor inductance	0	H
I_m	Moment of inertia	13.5e–7	kgm^2
g	Gravitational constant	9.81	m/s^2

Table 9.1: Parameter values for the actuated pendulum system

Each actuated pendulum P_i is described by a linearized continuous-time state equation

$$\underbrace{\begin{pmatrix} \dot\theta_i(t) \\ \ddot\theta_i(t) \end{pmatrix}}_{\dot{\boldsymbol{x}}_{ci}(t)} = \underbrace{\begin{pmatrix} 0 & 1 \\ \frac{(m_{ci}\ell_{ci}+m_{ei}\ell_i)g}{I_{pi}+k_r^2 I_m} & \frac{-k_r^2 k_t k_v}{(I_{pi}+k_r^2 I_m)R_a} \end{pmatrix}}_{\boldsymbol{A}_{ci}} \underbrace{\begin{pmatrix} \theta_i(t) \\ \dot\theta_i(t) \end{pmatrix}}_{\boldsymbol{x}_{ci}(t)} + \underbrace{\begin{pmatrix} 0 \\ \frac{k_r k_t}{(I_{pi}+k_r^2 I_m)R_a} \end{pmatrix}}_{\boldsymbol{b}_{ci}} \big(u_i(t)+d_i(t)\big) \tag{9.2}$$

where $\theta_i(t)$ is the pendulum angle in radian, $u_i(t)$ is the input voltage in volt, and $d_i(t)$ is the disturbance voltage in volt. Moreover, $\ell_{ci} = 0.5\ell_i$ is the pendulum length at center-point and $I_{pi} = m_{ci}\ell_{ci}^2 + m_{ei}\ell_i^2$ is the pendulum moment of inertia. The eigenvalues of the actuated pendulums are

$$\lambda_{11} = +00.76, \quad \lambda_{21} = +01.32, \quad \lambda_{31} = +02.28 \tag{9.3a}$$
$$\lambda_{12} = -50.63, \quad \lambda_{22} = -27.10, \quad \lambda_{32} = -11.96. \tag{9.3b}$$

The actuated pendulums are controlled over a single-channel communication network with the uncertain time-varying delays

$$\tau_i^{\mathrm{SC}}(k) \in [1\,\mathrm{ms}, 2\,\mathrm{ms}] \tag{9.4a}$$
$$\tau_i^{\mathrm{CA}}(k) \in [1\,\mathrm{ms}, 2\,\mathrm{ms}] \tag{9.4b}$$

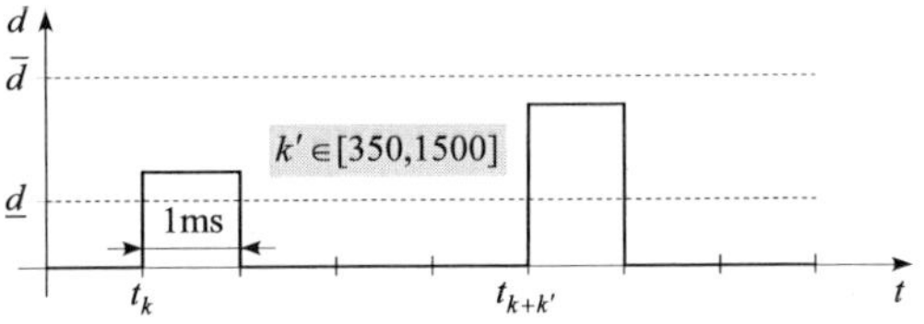

Figure 9.2: Model of an impulsive disturbance input

for all $i \in \mathbb{M}$. The weighting matrices of the continuous-time cost function (2.20) are finally chosen as

$$\boldsymbol{Q}_{ci} = \begin{pmatrix} 1000 & 0 \\ 0 & 1 \end{pmatrix}, \quad R_{ci} = 1 \quad \forall i \in \mathbb{M}. \tag{9.5}$$

Two simulation scenarios are performed in the following. In the first scenario, an initial value response is examined, i.e. $d_i(t) = 0\,\mathrm{V}$ for all $i \in \mathbb{M}$. A Monte Carlo simulation with one thousand random initial states $\boldsymbol{x}_{ci0} = \begin{pmatrix}\theta_{i0} & \dot{\theta}_{i0}\end{pmatrix}^T$, uniformly distributed over the unit interval $[0, 1]$, is performed under each codesign strategy. In the second scenario, a disturbance response is rather studied, i.e. $\boldsymbol{x}_{ci0} = \boldsymbol{0}$ for all $i \in \mathbb{M}$. A uniformly distributed impulsive disturbance voltage $d_i(t) \in [\underline{d}_i, \overline{d}_i]$, modeled in Figure 9.2 with $\underline{d}_i = -20\,\mathrm{V}$ and $\overline{d}_i = 20\,\mathrm{V}$, is considered. The duration of the disturbance impulse is equal to $T_{\text{imp}} = 1\,\mathrm{ms}$ despite the chosen sampling interval $h_{j(k)}$. In both scenarios, uniformly distributed random input delays $\tau_i(k) = \tau_i^{\mathrm{SC}}(k) + \tau_i^{\mathrm{CA}}(k)$, except for the PBCS strategy in which $\tau_i(k) = \tau_i^{\mathrm{CA}}(k)$, are considered. Moreover, the resulting cost is evaluated based on the cost function (2.29) for the simulation time $T_{\text{sim}} = 100\,\mathrm{s}$.

9.1.2 Simulation Results

Periodic Control and Scheduling

Networked control of the three actuated pendulums outlined above is first addressed using the PCS strategy. Starting from the simulation setup above, the parameters of the discrete-time switched polytopic system (2.18) and of the nominal cost function (4.4) are first obtained with the required design parameters chosen as

$$L = 5 \qquad \text{(Approximation order)} \tag{9.6a}$$
$$h_{j(k)} = h \in \{4\,\mathrm{ms}, 5\,\mathrm{ms}, \ldots, 10\,\mathrm{ms}\} \qquad \text{(Sampling interval)} \tag{9.6b}$$
$$\tau_{j(k)}(k) = \tau_{j(k)}^{\mathrm{SC}}(k) + \tau_{j(k)}^{\mathrm{CA}}(k) \in [2\,\mathrm{ms}, 4\,\mathrm{ms}] \qquad \text{(Input delay)} \tag{9.6c}$$
$$\tau_{0j(k)} = 3\,\mathrm{ms} \qquad \text{(Nominal input delay)} \tag{9.6d}$$

for all $j(k) \in \mathbb{M}$. Due to the network-induced input delays $\tau_{j(k)}(k) \in [2\,\mathrm{ms}, 4\,\mathrm{ms}]$, the smallest sampling interval that can be chosen while satisfying Assumption 2.2 is

$h = 4\,\text{ms}$. Based on the resulting NCS model and the associated nominal cost function, the set $\mathbb{S}_p$ of admissible p-periodic switching-feedback sequences σ defined in 4.3 is determined for different periods $p \in \{3, 4, 5, 6\}$ using Theorem 4.2. The resulting cardinality of $\mathbb{S}_p$ together with the computation time required for solving the LMI optimization problem (4.18) are similar to what obtained in Section 4.7 and summarized in Table 4.1.

Now, we are in the position to solve Problem 4.2. Three solutions to Problem 4.2 (with minor modifications) have been proposed in Chapter 4. The characteristics of each solution from the complexity and performance perspectives have been illustrated in more detail in Section 4.7. For redundancy avoidance, only the exhaustive search solution using Theorem 4.3 which leads to the best performance compared to the other solutions is considered in the following. The resulting average cost J_{avg} of the Monte Carlo simulation is depicted in Figure 9.3.

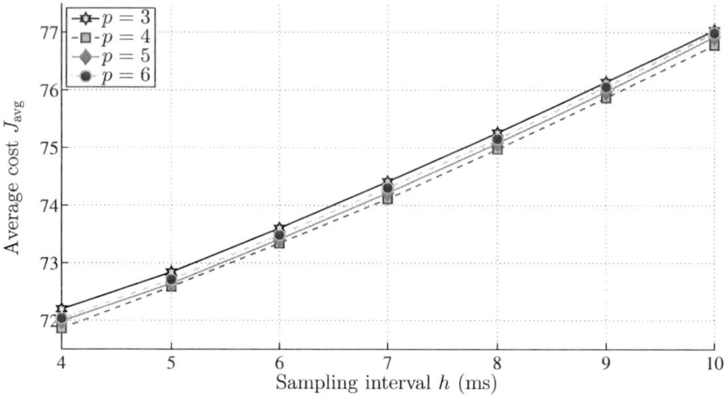

Figure 9.3: Resulting average cost versus sampling interval for the PCS strategy

Obviously, the average cost J_{avg} for each period p increases with the sampling interval h. A more interesting point to be observed is the cost behavior with respect to the period p. The change of the average cost with respect to p is quite marginal. Increasing p does not principally provide any benefit. It can even lead to less performance, e.g. between $p = 4$ and $p = \{5, 6\}$. This point has also been observed in the simulation results of the illustrative example in Section 4.7. This happens since for an initial value response all plants must be simultaneously controlled and hence a 3-periodic switching sequence such as $\big(j(0), j(1), j(2)\big) = (1, 2, 3)$ could be the optimal choice despite the chosen period p. To check whether this claim is right, the resulting switching sequence for different periods and a randomly chosen initial value

$$\begin{pmatrix}\theta_{10}\\ \dot{\theta}_{10}\end{pmatrix}=\begin{pmatrix}0.5881\\ 0.8977\end{pmatrix},\quad \begin{pmatrix}\theta_{20}\\ \dot{\theta}_{20}\end{pmatrix}=\begin{pmatrix}0.8915\\ 0.8158\end{pmatrix},\quad \begin{pmatrix}\theta_{30}\\ \dot{\theta}_{30}\end{pmatrix}=\begin{pmatrix}0.0359\\ 0.6918\end{pmatrix} \tag{9.7}$$

is illustrated in Figure 9.4. Note that for $t \in [0\,\mathrm{s}, 0.05\,\mathrm{s}]$, where the relevant dynamics is mostly covered, the resulting switching sequence is nothing but the 3-periodic switching sequence $(2, 1, 3)$, emphasizing the claim above.

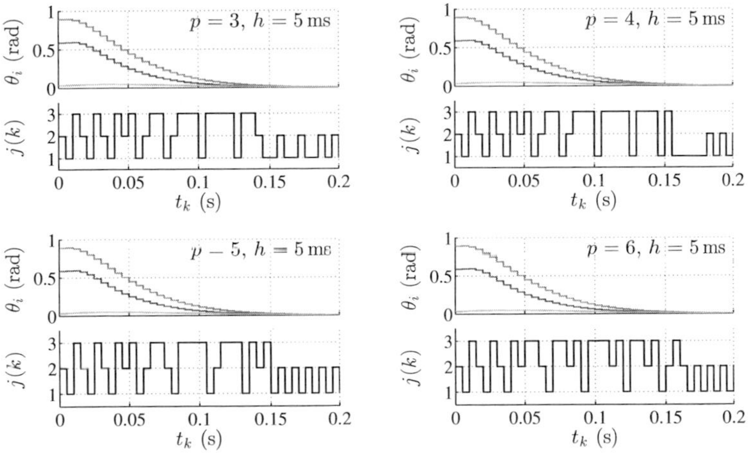

Figure 9.4: Pendulum angle θ_i and switching sequence $j(k)$ for different periods

The situation is different for a disturbance response in which different scenarios might occur within the specified simulation time, e.g. one plant is subject to a disturbance while the other plants not or two of them are simultaneously disturbed or even all of them and so on. Under such circumstances, increasing the period p which in turn increases the number of admissible p-periodic switching-feedback sequences σ might have a significant effect on the resulting performance. Simulation results for disturbance voltages $d_i(t)$, $\forall i \in \mathbb{M}$, are shown in Figure 9.5. Evidently, the larger the chosen value of p, the higher the resulting performance. This improvement of the performance is however at the cost of more offline/online computational complexity. Thus, the period p can be used as a design parameter to balance between performance and computational complexity.

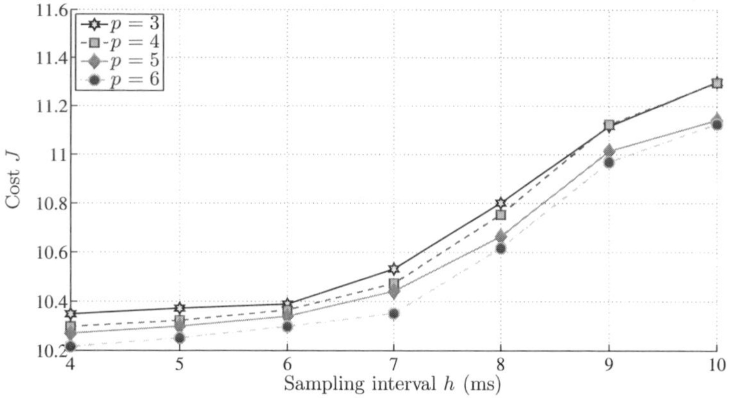

Figure 9.5: Resulting cost of disturbance rejection for the PCS strategy

Receding-Horizon Control and Scheduling

The RHCS strategy is now used for networked control of the three actuated pendulums outlined above. As for the PCS strategy, the parameters of the discrete-time switched polytopic system (2.18) and of the nominal cost function (5.4) are first obtained. The design parameters are chosen as in (9.6). The terminal weighting matrix $\boldsymbol{P}_{\sigma}(0)$ is determined based on Theorem 4.2 with the period $p = M$. For a fair comparison with the PCS strategy, different prediction horizons $N \in \{3, 4, 5, 6\}$ are chosen.

Based on the resulting NCS model and the associated nominal cost function, Problem 5.1 is then solved based on dynamic programming for each prediction horizon N using Theorem 5.1. The computation time required for solving the LMI optimization problem (5.21) is summarized in Table 9.2. The relaxed dynamic programming solution has been studied in Section 5.4 in more detail and hence it is not considered here.

Horizon N	No. of Branches $\lvert\mathbb{L}_0\rvert$	Computation Time
3	27	28.15 s
4	81	01.45 min
5	243	04.28 min
6	729	13.28 min

Table 9.2: Offline computational complexity for the RHCS strategy

The resulting average cost J_{avg} of the Monte Carlo simulation is depicted in Figure 9.6.

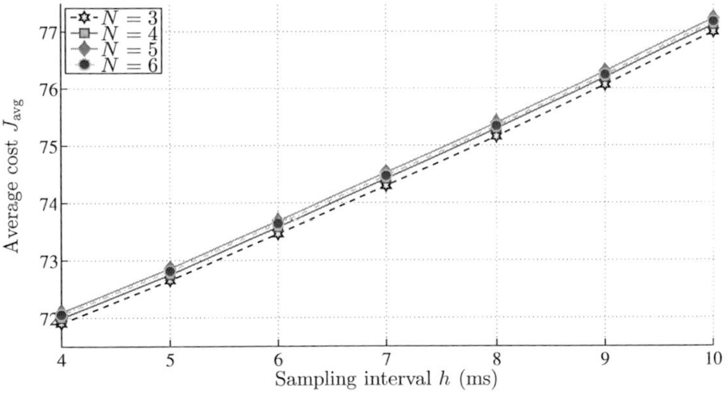

Figure 9.6: Resulting average cost versus sampling interval for the RHCS strategy

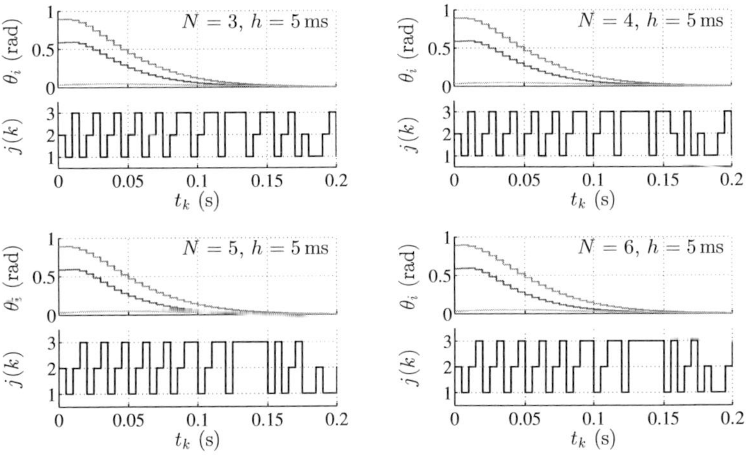

Figure 9.7: Pendulum angle θ_i and switching sequence $j(k)$ for different horizons

As for the PCS strategy, the average cost J_{avg} for each prediction horizon N increases with the sampling interval h. Furthermore, the change of the average cost with respect to the prediction horizon N is quite marginal. Increasing N does not principally provide any benefit. The reason behind this is the same reason outlined above for the PCS strategy. This can clearly be observed from the resulting switching sequence, depicted in Figure 9.7, for different prediction horizons and the random initial value given in (9.7). For $t \in [0\,\mathrm{s}, 0.1\,\mathrm{s}]$ where the relevant dynamics is mostly covered, the resulting switching sequence is the 3-periodic switching sequence $(1, 2, 3)$ despite the chosen prediction horizon N.

The situation is surely different for a disturbance response in which infinite realizations of the disturbance might occur within the simulation time since the disturbance release instant is random. Under such circumstances, increasing the prediction horizon N which in turn increases the number of branches in the resulting switching tree might have a significant effect on the resulting performance. The initial value response can thus be seen as a special case of the disturbance response, where all plants are subject to a disturbance impulse (initial value) at the same time instant t_0. Simulation results for impulsive disturbance voltages $d_i(t)$, $\forall i \in \mathbb{M}$, are illustrated in Figure 9.8.

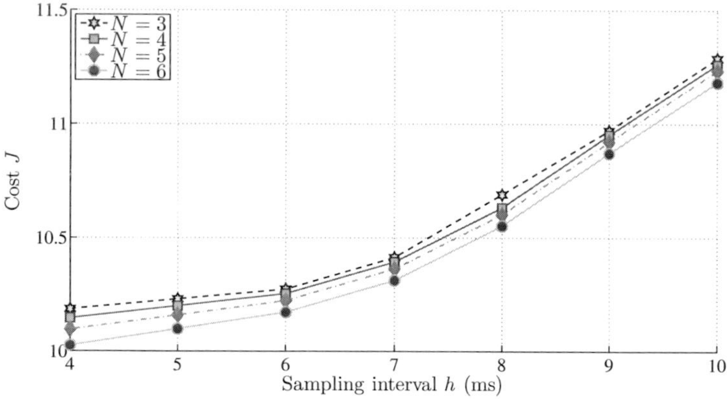

Figure 9.8: Resulting cost of disturbance rejection for the RHCS strategy

Obviously, the larger the chosen prediction horizon N, the lower the resulting cost. This improvement of the performance is at the cost of more computational complexity. Thus, the prediction horizon N can also be used as a design parameter for balancing between control performance and computational complexity.

Implementation-Aware Control and Scheduling

This time, the three actuated pendulums are controlled over the single-channel communication network using the IACS strategy. Following the same design procedure as for the PCS and the RHCS strategy, the parameters of the discrete-time switched polytopic system (2.18) and of the nominal cost function (6.4) are first obtained. The design parameters are chosen as in (9.6). Based on the resulting NCS model and the associated nominal cost function, Problem 6.1 is then solved online with the objective function (6.12a) using Algorithm 6.1. Problem 6.1 is further solved yet offline using Theorem 6.2 with objective function (6.12b). The Metzler matrix Ξ is chosen according to (6.19). The resulting average cost J_{avg} of the Monte Carlo simulation is depicted in Figure 9.9.

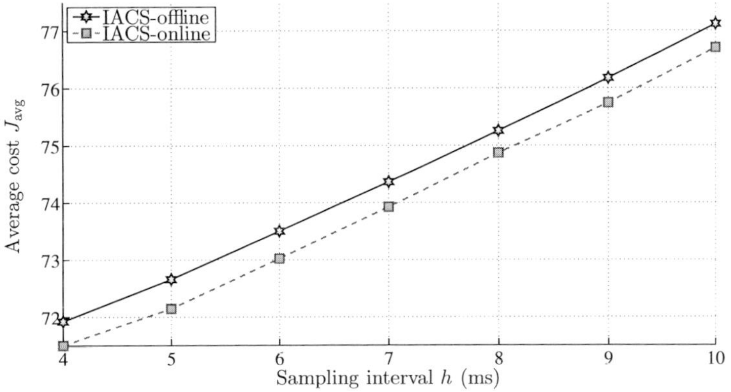

Figure 9.9: Resulting average cost versus sampling interval for the IACS strategy

As for the PCS/RHCS strategy, the average cost J_{avg} increases with the sampling interval h. Moreover, the resulting average cost for the online IACS strategy using Algorithm 6.1 is lower than the offline counterpart. This improvement in the control performance is due to the reoptimization of the control parameters $\boldsymbol{K}_{j(k)}$ and the scheduling parameters $\boldsymbol{P}_{j(k)}$ according to the current state $\boldsymbol{x}(k)$, reducing the conservatism of the robust IACS strategy. Simulation results for impulsive disturbance voltages $d_i(t)$, $\forall i \in \mathbb{M}$, are shown in Figure 9.10. It is quite obvious that the reoptimization of the control and scheduling parameters for the current state measurements significantly improves the resulting performance compared to the permanent utilization of the offline computed parameters.

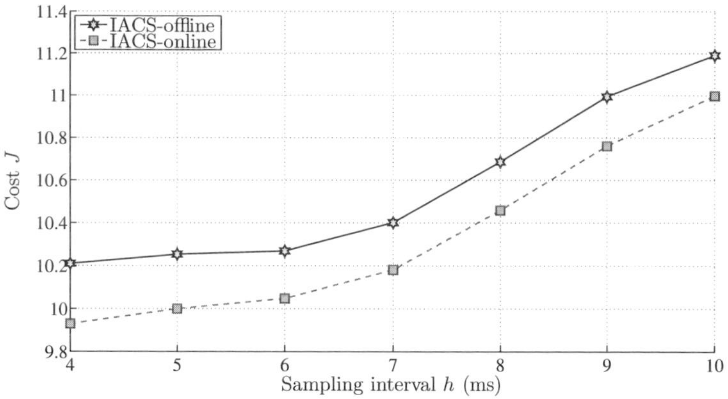

Figure 9.10: Resulting cost of disturbance rejection for the IACS strategy

Event-Based Control and Scheduling

Networked control of the three actuated pendulums is penultimately addressed using the EBCS strategy. For the simulation setup given above, the parameters of the discrete-time switched polytopic system (7.14) and of the nominal cost function (7.18) are first obtained. The required design parameters are chosen according to (9.6) except the sampling interval $h_{j(k)}$ which is chosen as

$$h_{j(k)} = 4\,\text{ms} \qquad \forall j(k) \in \mathbb{M}_0. \tag{9.8}$$

Note that due to the network-induced input delays $\tau_{j(k)}(k) \in [2\,\text{ms}, 4\,\text{ms}]$ the chosen value in (9.8) is the smallest sampling interval that can be chosen while satisfying Assumption 2.2. Based on the resulting NCS model and the associated nominal cost function, Problem 7.1 is then solved for different tuning parameters $\lambda_i = \lambda \in \{0.2, 0.25, \ldots, 0.75\}$ based on the S-procedure using Theorem 7.1. The resulting average cost J_{avg} of the Monte Carlo simulation is depicted in Figure 9.11. It apparently increases with the tuning parameter λ. The higher the chosen value of λ, the lower the utilization of the communication network (as deduced from the inter-event time) and the lower the resulting control performance. For $\lambda > 0.75$, the LMI optimization problem (7.35) in Theorem 7.1 is not feasible anymore. The same conclusion can be made for the disturbance response, as depicted in Figure 9.12. Thus, the control performance and network utilization can systematically be balanced via the tuning parameter λ.

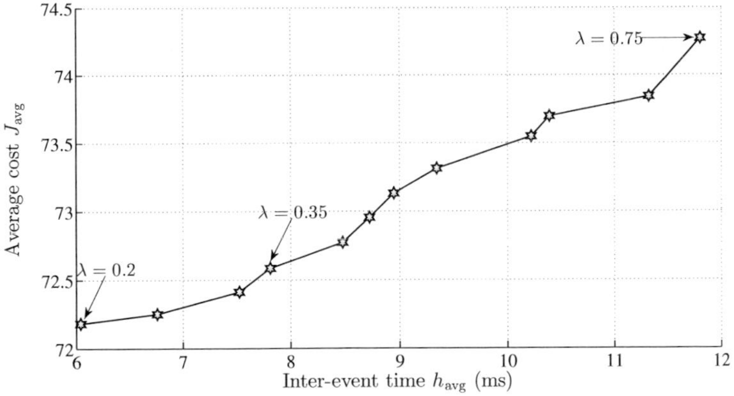

Figure 9.11: Resulting average cost versus inter-event time for the EBCS strategy

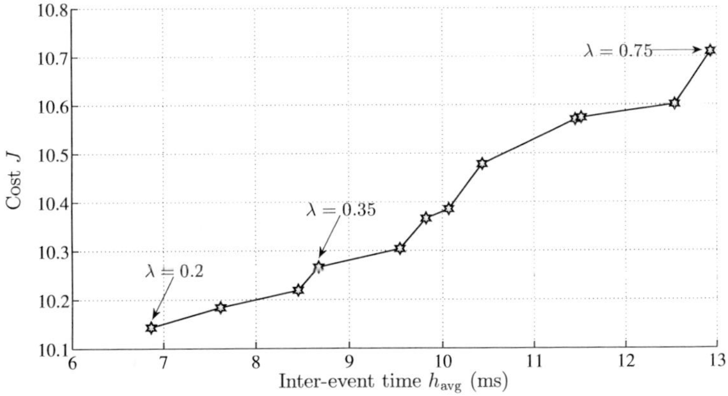

Figure 9.12: Resulting cost of disturbance rejection for the EBCS strategy

A sketch of the disturbance response of the three actuated pendulums for the first eight seconds, i.e. $t \in [0\,\text{s}, 8\,\text{s}]$, and the tuning parameter $\lambda = 0.001$ is shown in Figure 9.13.

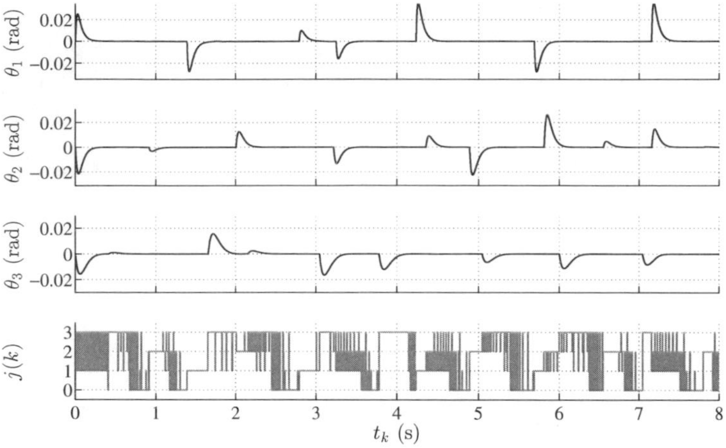

Figure 9.13: Disturbance response for the EBCS strategy with $\lambda = 0.001$

Apparently, the network access is granted to the more demanding plant as seen from the output $j(k)$ of the scheduler. For $t \in [0.0\,\text{s}, 0.4\,\text{s}]$, all plants for example are alternately granted access since all of them are disturbed. For $t \in [0.9\,\text{s}, 1.1\,\text{s}]$, however, $j(k) = 2$ is decided since only P_2 is subject to a disturbance impulse. The same behavior results for P_1 within $t \in [1.4\,\text{s}, 1.65\,\text{s}]$ and for P_3 within $t \in [3.8\,\text{s}, 4.15\,\text{s}]$. For $t \in [0.67\,\text{s}, 0.92\,\text{s}]$ and $t \in [2.45\,\text{s}, 2.8\,\text{s}]$, the idle task T_0 is preferably chosen since all P_i are very close to the equilibrium and hence no necessity for control exists. Thus, an efficient distribution and utilization of the limited communication resource is indeed guaranteed under the proposed EBCS strategy.

Prediction-Based Control and Scheduling

Ultimately, networked control of the three actuated pendulums is addressed using the PBCS strategy. For the simulation setup given above, the parameters of the discrete-time switched polytopic system (8.13) and of the nominal cost function (8.17) are first obtained. The required design parameters are chosen as

$$L = 5 \tag{9.9a}$$

$$h_{j(k)} = 2\,\text{ms} \quad \forall j(k) \in \mathbb{M}_0 \tag{9.9b}$$

$$\tau(k) = \tau^{\text{CA}}(k) \in [1\,\text{ms}, 2\,\text{ms}] \tag{9.9c}$$

$$\tau_0 = 1.5\,\text{ms}. \tag{9.9d}$$

Note that the network-induced input delay $\tau(k)$ given in (9.9) differs from the one given in (9.6). Only $\tau^{\mathrm{CA}}(k)$ that is required for transmitting new control updates to corresponding plants is considered while $\tau^{\mathrm{SC}}_{j(k)}(k)$ required for transmitting new state measurements is compensated by the model-based predictor introduced within each control loop, cf. Remark 8.1. Thus, the smallest sampling interval that can be chosen while satisfying Assumption 2.2 and the relationship in (8.4) is $h_{j(k)} = 2\,\mathrm{ms}$. Based on the resulting NCS model and the associated nominal cost function, Problem 8.1 is then solved for different $\lambda_{\mathrm{x}i} = \lambda_{\mathrm{x}} \in \{0.001, 0.01, 0.1\}$ and $\lambda_{\mathrm{u}} \in \{0.0, 0.05, \ldots, \lambda_{\mathrm{u}}^{\max}\}$ based on the S-procedure using Theorem 8.1. The resulting average cost J_{avg} of the Monte Carlo simulation is depicted in Figure 9.14.

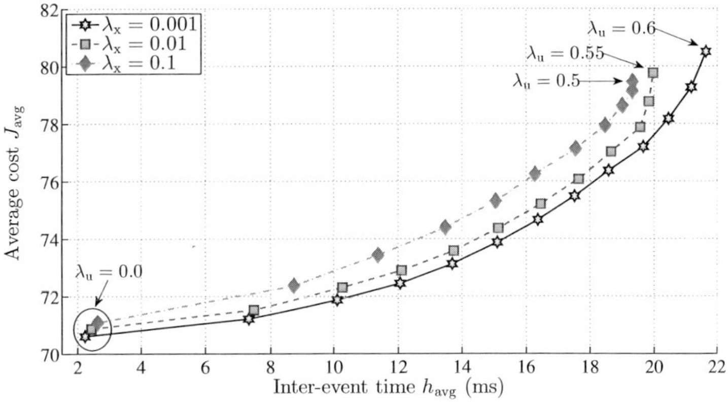

Figure 9.14: Resulting average cost versus inter-event time for the PBCS strategy

Similar to the simulation results printed in Section 8.4, the average cost J_{avg} increases with the tuning parameters λ_{x} and λ_{u}. The higher the chosen values of λ_{x} and/or λ_{u}, the lower the utilization of the communication network (as deduced from the inter-event time) and thus the lower the resulting control performance. The same conclusion can be made for the disturbance response, as depicted in Figure 9.15. Thus, the control performance and network utilization can systematically be balanced via the tuning parameters λ_{x} and λ_{u}.

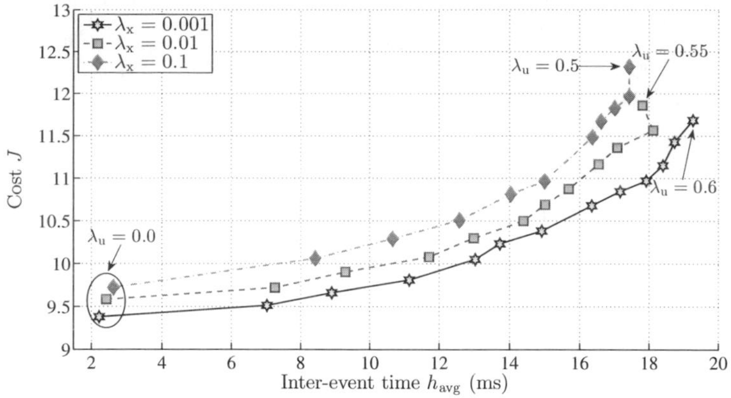

Figure 9.15: Resulting cost of disturbance rejection for the PBCS strategy

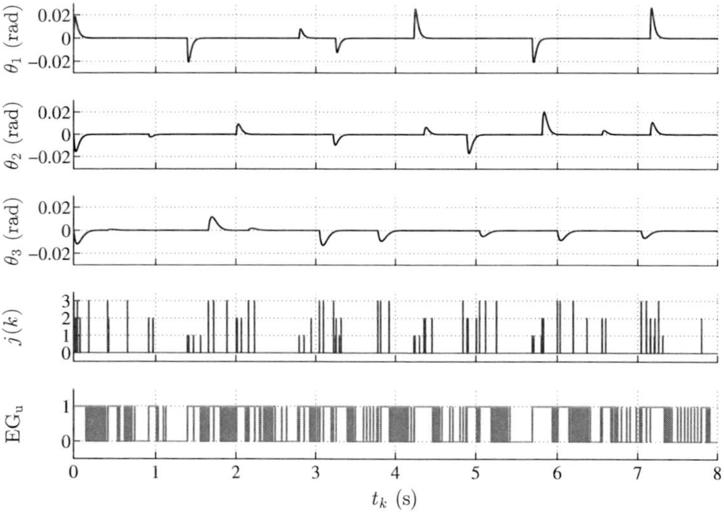

Figure 9.16: Disturbance response for the PBCS strategy with $\lambda_x = \lambda_u = 0.001$

A sketch of the disturbance response of the three actuated pendulums for the first eight seconds, i.e. $t \in [0\,\mathrm{s}, 8\,\mathrm{s}]$, with the tuning parameters $\lambda_\mathrm{x} = \lambda_\mathrm{u} = 0.001$ is shown in Figure 9.16. It can easily be seen that the output of the scheduler is decided according to the demands of the plants. For instance, the output of the scheduler is $j(k) = 2$ once the plant P_2 is subject to a disturbance impulse at time instant $t = 0.92\,\mathrm{s}$. The same behavior results for P_1 at time instant $t = 1.39\,\mathrm{s}$ and for P_3 at time instant $t = 1.66\,\mathrm{s}$. For the event generator EG_u at the controller side, the same behavior can be observed with the difference that all control signals are updated in case of an event is triggered. Overall, an efficient distribution and utilization of the limited communication resource is guaranteed under the proposed PBCS strategy as well.

9.1.3 Conclusions

The three actuated pendulums outlined above are controlled over a single-channel communication network using the five control and scheduling codesign strategies proposed in this thesis. The simulation results of each are given in the previous section. In this section, we compare between the codesign strategies from the control performance perspective. To this end, the best of the achieved performance under each codesign strategy is considered in the following.

For the first simulation scenario, in which a Monte Carlo simulation with one thousand uniformly distributed random initial states is performed, the resulting average cost for each codesign strategy is depicted in Figure 9.17.

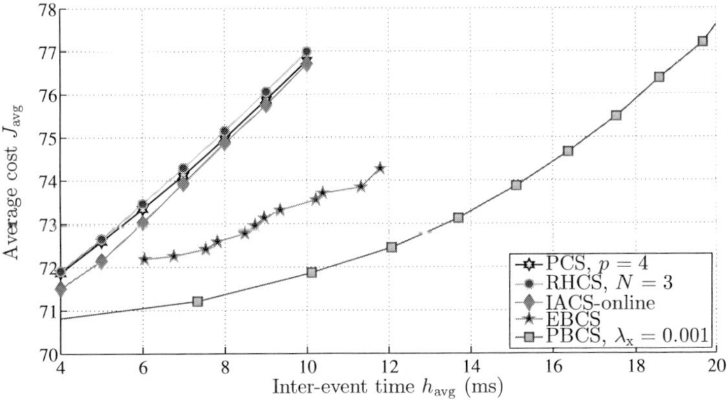

Figure 9.17: Resulting mean cost versus inter-event time for different codesign strategies

Although the aim of the RHCS strategy is to reduce the conservatism of the PCS strategy due to the imposed periodicity, it leads for initial value response to higher average cost than the PCS strategy. Going back to Figure 9.4 or Figure 9.7, however, the reason behind such performance degradation based on the RHCS strategy can easily be explained. This is because of the strong demand of all plants to be simultaneously controlled. Hence, any of the p-periodic switching sequences explicitly considered in the PCS strategy could lead to better performance than the generally non-periodic switching sequences of the RHCS strategy.

The IACS strategy leads to a smaller average cost than the PCS/RHCS strategy. This improvement is attributed to the online optimization of the feedback and scheduling matrices according to the current state $\boldsymbol{x}(k)$ and not completely offline as for the PCS and RHCS strategy. In other words, considering different periods for the PCS strategy or different prediction horizons for the RHCS strategy can not compensate the conservatism induced by the state-independent design of the control and scheduling parameters.

The EBCS strategy leads principally to a considerably smaller cost than the other time-triggered codesign strategies. The smallest cost is obtained based on the PBCS strategy which is attributed to the ability of simultaneously controlling all plants based on the available information about system dynamics on the associated event generators as well as on the centralized controller.

For the second simulation scenario, in which the response for uniformly distributed impulsive disturbance inputs and zero initial states is studied, the resulting cost of disturbance rejection for each codesign strategy is depicted in Figure 9.18.

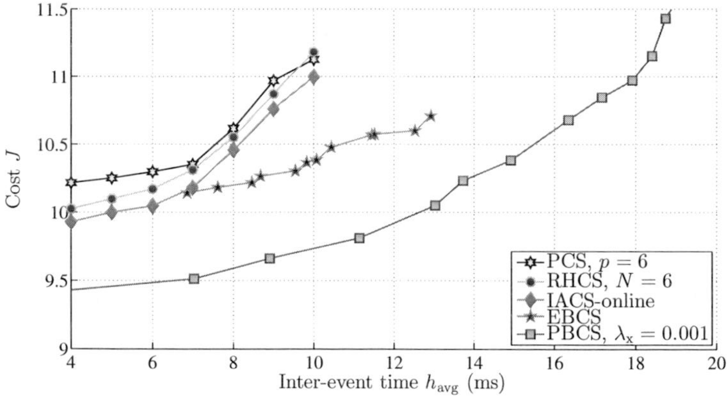

Figure 9.18: Resulting cost of disturbance rejection for different codesign strategies

The same conclusion as for the first simulation scenario can also be made except that the RHCS strategy principally outperforms the PCS strategy. This is to be expected since the number of switching sequence candidates to be inspected for the RHCS strategy is much higher than that for the PCS strategy, giving more freedom in choosing the best switching sequence for each realization of the disturbance.

9.2 Implementation

9.2.1 Experimental Setup

Consider a networked control of two actuated pendulums, as illustrated in Figure 9.19. Each actuated pendulum P_i is described by the linearized continuous-time state equation (9.2) with the parameters given according to (9.1) and Table 9.1.

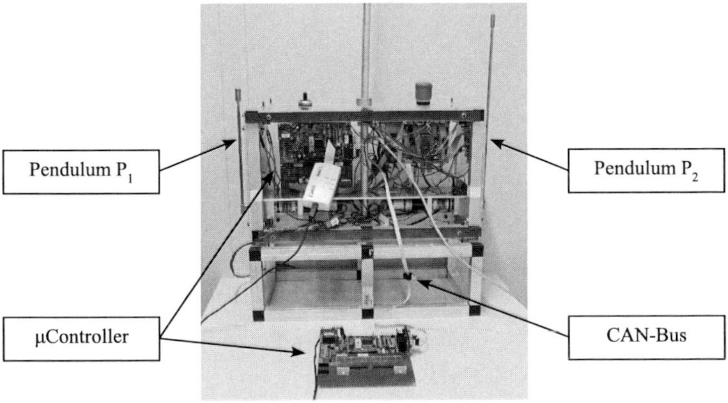

Figure 9.19: Two actuated pendulums controlled over a CAN bus

Consider further the continuous-time cost function (2.20) with the weighting matrices chosen as in (9.5). The two actuated pendulums are controlled over a CAN bus (bit rate 500 kbit/s and cable length 69 cm) with the artificially induced (as a random waiting time) uncertain time-varying delays

$$\tau_i^{\mathrm{SC}}(k) \in [2\,\mathrm{ms}, 4\,\mathrm{ms}] \tag{9.10a}$$

$$\tau_i^{\mathrm{CA}}(k) \in [2\,\mathrm{ms}, 4\,\mathrm{ms}] \tag{9.10b}$$

for all $i \in \mathbb{M} = \{1, 2\}$. Furthermore, two microcontrollers (32-bit phyCORE-LPC2294, ARM7-based CPU, 60 MHz clock frequency) are used for the implementation of the scheduling algorithm and the control algorithm of each codesign strategy. The communication between the microcontrollers is done over the CAN bus as seen in Figure 9.19.

The computational delays τ_{sch} and τ_{cont} required for respectively executing the scheduling algorithm and the control algorithm based on the given platform are indicated in the next section for each codesign strategy. It is worth to note that the computational delay τ_{sch} subsumes also the time delay required for reading the required state measurements using a serial peripheral interface (SPI) bus.

The experiment scenario is as follows: Each of the proposed codesign strategies is tested for uniformly distributed impulsive disturbance voltages $d_i(t) \in [-20\,\text{V}, 20\,\text{V}]$, characterized in Figure 9.2, and zero initial states $\boldsymbol{x}_{ci0} = \boldsymbol{0}$ for all $i \in \mathbb{M}$. Moreover, a uniformly distributed random input delay

$$\tau_i(k) = \tau_{\text{sch}} + \tau_i^{\text{SC}}(k) + \tau_{\text{cont}} + \tau_i^{\text{CA}}(k) \tag{9.11}$$

is considered for each codesign strategy. Exception is the PBCS strategy in which the associated input delay is given by

$$\tau_i(k) = \tau_{\text{cont}} + \tau_i^{\text{CA}}(k). \tag{9.12}$$

The resulting cost is finally evaluated based on the cost function (2.29) for the experimental time $T_{\text{exp}} = 180\,\text{s}$.

9.2.2 Experimental Results

Periodic Control and Scheduling

For less online computational complexity and thus less computational delay, the solution of Problem 4.2 based on relaxation using Theorem 4.4 is considered in the following. The required design parameters are summarized in Table 9.3.

Parameter	Description	Value	Unit
p	Sequence period	4	–
α	Relaxation factor	2	–
L	Approximation order	5	–
$h_{j(k)}$	Sampling interval	40	ms
$\tau_{j(k)}$	Random input delay	$[10, 15]$	ms
$\tau_{0j(k)}$	Nominal input delay	12.5	ms

Table 9.3: Parameter values for the PCS strategy

The resulting cardinality of the relaxed set $\hat{\mathbb{S}}_p$, the computation time τ_{sch} of the region membership test (4.41), the computation time τ_{cont} of the control algorithm (3.1), and the resulting cost J of disturbance rejection are indicated in Table 9.4.

Strategy	$\lvert\hat{\mathbb{S}}_p\rvert$	τ_{sch}	τ_{cont}	J
PCS ($\alpha = 2$)	6	6.62 ms	0.07 ms	56.78

Table 9.4: Experimental results for the PCS strategy

A sketch of the disturbance response of the two actuated pendulums for the first thirty-five seconds, i.e. $t \in [0\,\text{s}, 35\,\text{s}]$, is shown in Figure 9.20. It can be seen that the scheduler reacts directly to any disturbance impulse acting on any plant P_i by allowing its current state measurement to be sent for further execution of the corresponding control task T_i. In case that none of the plants is disturbed, one of them is randomly chosen for control.

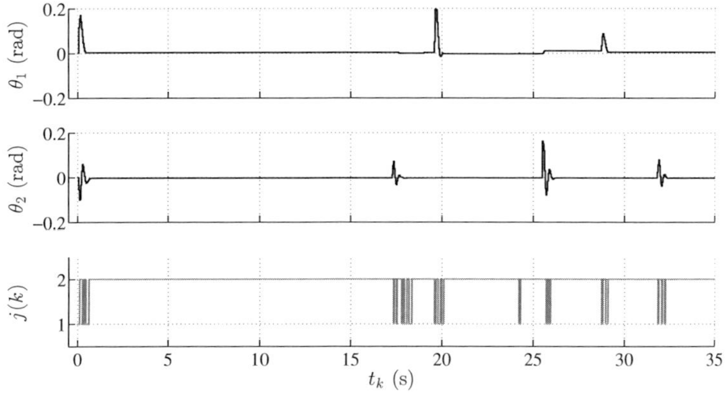

Figure 9.20: Disturbance response for the PCS strategy

Receding-Horizon Control and Scheduling

As for the PCS strategy, the relaxed dynamic programming solution to Problem 5.1 using Algorithm 5.1 is considered here for less online computational complexity. The required design parameters are summarized in Table 9.5.

Parameter	Description	Value	Unit
N	Prediction horizon	4	–
α	Relaxation factor	2	–
L	Approximation order	5	–
$h_{j(k)}$	Sampling interval	40	ms
$\tau_{j(k)}$	Random input delay	$[10, 15]$	ms
$\tau_{0j(k)}$	Nominal input delay	12.5	ms

Table 9.5: Parameter values for the RHCS strategy

The resulting cardinality of the relaxed set $\hat{\mathbb{L}}_0$, the computation time τ_{sch} of the region membership test (5.10), the computation time τ_{cont} of the control algorithm (3.1), and the resulting cost J of disturbance rejection are indicated in Table 9.6.

Strategy	$\lvert\hat{\mathbb{L}}_0\rvert$	τ_{sch}	τ_{cont}	J
RHCS ($\alpha = 2$)	6	6.62 ms	0.07 ms	51.24

Table 9.6: Experimental results for the RHCS strategy

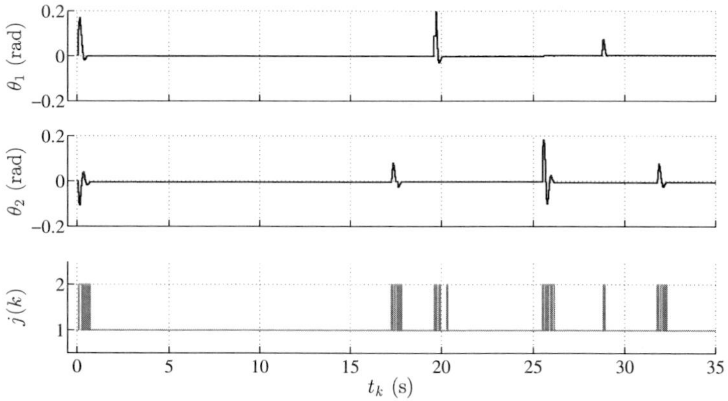

Figure 9.21: Disturbance response for the RHCS strategy

A sketch of the disturbance response of the two actuated pendulums for the first thirty-five seconds, i.e. $t \in [0\,\text{s}, 35\,\text{s}]$, is shown in Figure 9.21. As for the PCS strategy, one can notice that the scheduler reacts directly to any disturbance impulse acting on any plant P_i by allowing its current state measurement to be sent for further execution of the corresponding control task T_i. Partly control tasks are executed though the related pendulums are not disturbed, e.g. around $t_k = 20\,\mathrm{s}$ where the second plant P_2 is controlled as well. This is because of some hardware problems, particularly in the drive unit, which lead to kind of coupling between the pendulums such that disturbing one of them affects the other one by somehow. This effect is beyond the scope of the experiment and thus it is not further investigated.

Implementation-Aware Control and Scheduling

We consider here for less online computational complexity only the offline solution to Problem 6.1 using Theorem 6.2 with the objective function (6.12b). The Metzler matrix Ξ is chosen according to (6.19). The required design parameters are summarized in Table 9.7.

Parameter	Description	Value	Unit
L	Approximation order	5	–
$h_{j(k)}$	Sampling interval	40	ms
$\tau_{j(k)}$	Random input delay	$[7, 12]$	ms
$\tau_{0j(k)}$	Nominal input delay	9.5	ms

Table 9.7: Parameter values for the IACS strategy

The computation time τ_{sch} of the switching law (6.6), the computation time τ_{cont} of the control algorithm (3.1), and the resulting cost J of disturbance rejection are indicated in Table 9.6.

Strategy	τ_{sch}	τ_{cont}	J
IACS-offline	3.13 ms	0.07 ms	59.58

Table 9.8: Experimental results for the IACS strategy

A sketch of the disturbance response of the two actuated pendulums for the first thirty-five seconds, i.e. $t \in [0\,\mathrm{s}, 35\,\mathrm{s}]$, is shown in Figure 9.21. The same conclusion as for the PCS/RHCS strategy can be made here as well.

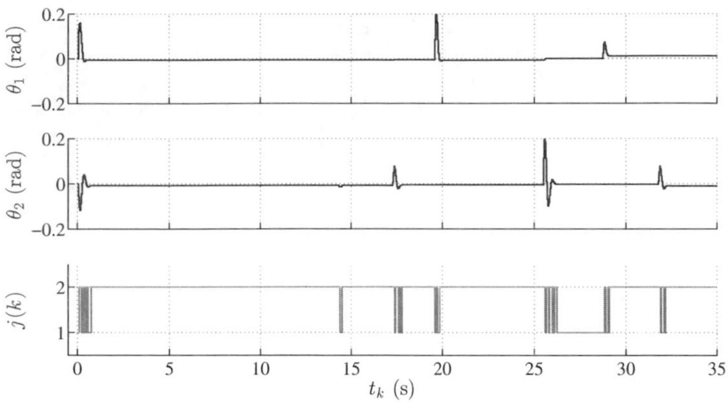

Figure 9.22: Disturbance response for the IACS strategy

Event-Based Control and Scheduling

Now, Problem 7.1 is solved based on the S-procedure using Theorem 7.1 with the design parameters indicated in Table 9.9.

Parameter	Description	Value	Unit
λ_i	Tuning parameter	0.1	–
L	Approximation order	5	–
$h_{j(k)}$	Sampling interval	20	ms
$\tau_{j(k)}$	Random input delay	$[5, 10]$	ms
$\tau_{0j(k)}$	Nominal input delay	7.5	ms

Table 9.9: Parameter values for the EBCS strategy

The computation time τ_{sch} of the switching law (7.22), the computation time τ_{cont} of the control algorithm (7.2), and the resulting cost J of disturbance rejection are indicated in Table 9.10.

Strategy	τ_{sch}	τ_{cont}	Update rate	J
EBCS	1.8 ms	0.06 ms	4%	45.03

Table 9.10: Experimental results for the EBCS strategy

A sketch of the disturbance response of the two actuated pendulums for the first thirty-five seconds, i.e. $t \in [0\,\mathrm{s}, 35\,\mathrm{s}]$, is shown in Figure 9.23. The network access is apparently granted to the more demanding plant, as seen from the output $j(k)$ of the scheduler. Once no necessity for control exists, moreover, the idle task T_0 is chosen for an efficient utilization of the limited resource.

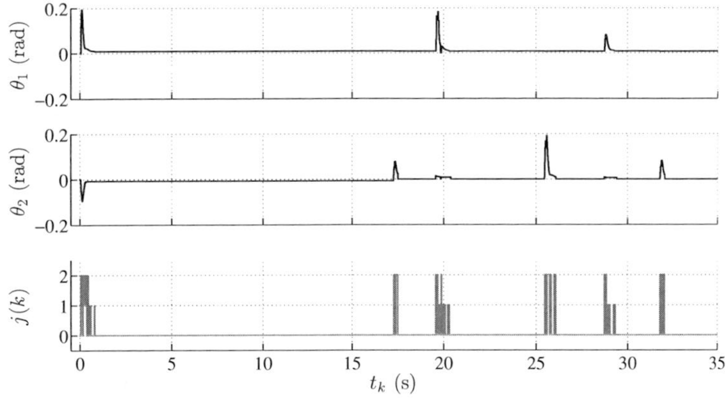

Figure 9.23: Disturbance response for the EBCS strategy

Prediction-Based Control and Scheduling

Ultimately, Problem 8.1 is solved based on the S-procedure using Theorem 8.1 with the design parameters indicated in Table 9.11.

Parameter	Description	Value	Unit
$\lambda_{\varkappa i}$	Tuning parameter	0.1	–
λ_{u}	Tuning parameter	0.1	–
L	Approximation order	5	–
$h_{j(k)}$	Sampling interval	20	ms
$\tau(k)$	Random input delay	$[4, 7]$	ms
$\tau_{0j(k)}$	Nominal input delay	5.5	ms

Table 9.11: Parameter values for the PBCS strategy

It is worth to note that induced input delay $\tau(k)$ above subsumes only τ_{cont} and $\tau^{\mathrm{CA}}(k)$, as shown in (9.12). The other induced delays such as the scheduling time τ_{sch} and the transmission time $\tau^{\mathrm{SC}}_{j(k)}(k)$ are compensated by the utilized mode-based predictors. The computation time τ_{sch} of the switching law (8.21), the computation time τ_{cont} of the control algorithm (8.10), and the resulting cost J of disturbance rejection are indicated in Table 9.12.

Strategy	τ_{sch}	τ_{cont}	Update rate	J
PBCS	2.3 ms	1.9 ms	29%	46.53

Table 9.12: Experimental results for the PBCS strategy

A sketch of the disturbance response of the two actuated pendulums for the first thirty-five seconds, i.e. $t \in [0\,\text{s}, 35\,\text{s}]$, is shown in Figure 9.24. The same conclusion as for the EBCS strategy can be made except for $t \in [25.5\,\text{s}, 32.5\,\text{s}]$ in which $j(k) = 2$ is preferably decided due to the actuator dead-zone that is not modeled in the corresponding model-based predictor.

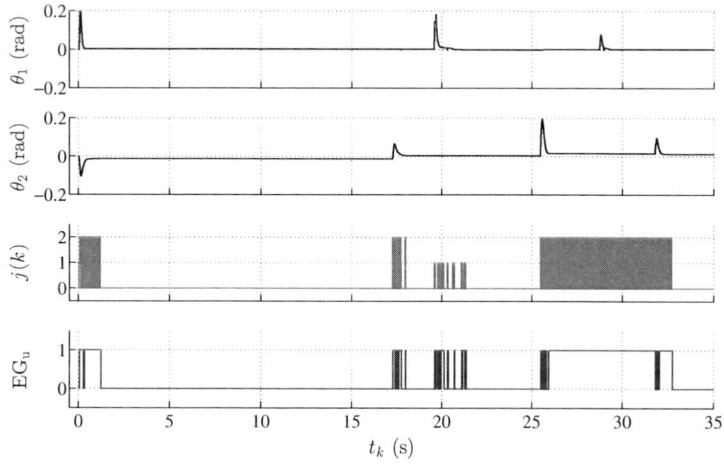

Figure 9.24: Disturbance response for the PBCS strategy

9.2.3 Conclusions

The control and scheduling codesign strategies proposed in Chapters 4–8 are implemented for networked control of the two actuated pendulums outlined in the experimental setup above. The main experimental results of each are summarized in Table 9.13.

Strategy	Sampling interval $h_{j(k)}$	Update rate	Inter-event time h_{avg}	Cost J
TDMA	40 ms	100%	40 ms	68.64
IACS-offline	40 ms	100%	40 ms	59.58
PCS ($p = 4$, $\alpha = 2$)	40 ms	100%	40 ms	56.78
RHCS ($N = 4$, $\alpha = 2$)	40 ms	100%	40 ms	51.24
PBCS ($\lambda_{xi} = \lambda_u = 0.1$)	20 ms	29%	68.96 ms	46.53
EBCS ($\lambda_i = 0.1$)	20 ms	4%	500 ms	45.03

Table 9.13: Comparison of the control and scheduling codesign strategies

For comparison purposes, the well-known TDMA scheduling strategy outlined in Section 1.4 with the 2-periodic switching-feedback sequence σ_{off} determined by (4.50) is also implemented. The computation time τ_{sch} including the state measuring process, the computation time τ_{cont} of the control algorithm (3.1), and the resulting cost J of disturbance rejection are indicated in Table 9.14. Note that the codesign strategies listed in Table 9.13 are sorted for a descending order of the resulting costs.

Strategy	τ_{sch}	τ_{cont}	J
TDMA	0.7 ms	0.07 ms	68.64

Table 9.14: Experimental results for the TDMA strategy

The conventional TDMA strategy leads to the largest cost. This is to be expected due to its static property where the predefined 2-periodic switching-feedback sequence σ_{off} must be followed at runtime despite the system states and demands.

A relatively better performance compared to the TDMA strategy has been obtained based on the offline version of the IACS strategy. This improvement in the resulting control performance is due to the efficient distribution of the limited communication resource according to the demands of the actuated pendulums.

The performance has further been improved based on the PCS strategy. This improvement is because of the larger number of switching sequence candidates to be inspected, in contrast to the IACS strategy where the pointer of one 2-periodic switching-feedback sequence, cf. Remark 6.4, is to be inspected. This gives in turn more freedom in choosing the best switching sequence for each realization of the disturbance.

More improvement of the control performance has been obtained using the RHCS strategy. This is attributed to the larger number of switching sequence candidates to be inspected compared to the PCS strategy. This fact has also been emphasized by the simulation results printed above.

The EBCS strategy leads surprisingly to the smallest cost and a huge inter-event time

h_{avg} compared to the other strategies including the PBCS strategy. Although the simulation results have shown that the PBCS strategy is much better than the EBCS strategy from the performance perspective, the experimental results show completely the opposite. The main reason behind the contradiction among the simulation and experimental results can however be easily explained. It is due to the actuator dead-zone that is not modeled in the associated model-based predictors, on the one hand, and the estimate of the non-measured state $\dot{\theta}_i$, $\forall i$, on the other hand. The dead-zone effect leads to high update rate, as shown in Figure 9.24 for $t \in [25.5\,\text{s}, 32.5\,\text{s}]$, while the estimation error of the non-measured states leads to a larger cost. These effects do not appear in the simulation results, leading thus to such contradiction.

10 Conclusions and Future Work

10.1 Conclusions

The problem of jointly designing a controller and a scheduler for networked control systems is addressed in this thesis. The NCS composing of multiple plants that share a common communication network with uncertain time-varying transmission times is modeled as a switched polytopic system with additive norm-bounded uncertainty. Switching is deployed to represent scheduling, the polytopic uncertainty to overapproximatively describe the uncertain time-varying transmission time. Based on the resulting NCS model and a state feedback control law, the codesign problem of the switching sequence $j(0), \ldots, j(\infty)$ and the feedback sequence $\boldsymbol{K}(0), \ldots, \boldsymbol{K}(\infty)$ is formulated as an optimization problem with the objective of minimizing the worst-case value of an infinite time-horizon quadratic cost function. Five robust codesign methods are proposed in the thesis for tackling the introduced optimization problem, each of them is shortly described below. The properties of each are evaluated and compared in terms of performance based on simulation and experimental study, showing their effectiveness in improving the control performance while utilizing the limited resources efficiently.

Periodic Control and Scheduling. For a tractable optimization problem, the switching sequence $j(0), \ldots, j(\infty)$ and thus the feedback sequence $\boldsymbol{K}(0), \ldots, \boldsymbol{K}(\infty)$ are assumed to be p-periodic, i.e. $j(k+p) = j(k)$ and $\boldsymbol{K}(k+p) = \boldsymbol{K}(k)$. Based on this assumption, the main codesign problem is transformed into the problem of determining the optimal p-periodic switching-feedback sequence σ^* such that the associated cost function is robustly minimized. This problem is solved in two steps. In the first step, the set $\mathbb{S}_p$ of admissible p-periodic switching-feedback sequences σ is determined. Admissibility of a p-periodic switching-feedback sequence σ is in the sense of the capability of stabilizing the switched polytopic system under consideration. Since the switching index $j(k)$ belongs to the finite set $\mathbb{M}$, all p-periodic switching sequences can be determined by permutations with repetition. The corresponding p-periodic feedback matrices $\boldsymbol{K}(k)$ are determined from an LMI optimization problem using p-periodic Lyapunov functions. Once the set $\mathbb{S}_p$ is determined, we run the second step by getting through all of the p-periodic switching-feedback sequence candidates $\sigma \in \mathbb{S}_p$, i.e. exhaustive search, and checking whether each candidate satisfies the problem's statement. In order to suppress the imposed periodicity and hence improve the resulting performance, the exhaustive search is performed at each time instant t_k for the current state $\boldsymbol{x}(k)$. For reducing the online complexity of the exhaustive search while introducing further suboptimality, two relaxed versions of the periodic codesign strategy are proposed. The effectiveness of the proposed strategy as well as the effect of the chosen period p on the resulting control performance are evaluated by simulation for an illustrative example and also experimentally for a case study.

Receding-Horizon Control and Scheduling. Instead of imposing periodicity on the complete switching and feedback sequences, the infinite time-horizon quadratic cost function is decomposed into two parts, J_1 and J_2. In the second part J_2, a stabilizing p-periodic switching-feedback sequence σ is considered. Based on this setup, the infinite time-horizon cost function (2.30) can be written as a finite time-horizon cost function J_N with the prediction horizon N. The main codesign problem can now be solved in a receding-horizon fashion based on dynamic programming, taking the current state $\boldsymbol{x}(k)$ explicitly into account. Although a major computational part of the dynamic programming solution can be performed offline, both offline and online complexity grow exponentially with the prediction horizon N. Such complexity can become computationally intractable (offline) or prohibitive (online) in some practical applications. The relaxed dynamic programming solution is therefore proposed to cope with the computational complexity. The resulting computational complexity can thus be tuned at the cost of relaxing the control performance to be achieved. The effectiveness of the proposed strategy and the effect of the prediction horizon N on the resulting performance are evaluated simulatively for an illustrative example and experimentally for a case study.

Implementation-Aware Control and Scheduling. In this strategy, neither imposed periodicity on the switching sequence nor partitioning of the cost function is considered. Instead, a state-based switching law with a quadratic form structure is introduced. Based on the switching law, the main codesign problem is formulated as an LMI optimization problem using Lyapunov-Metzler functions. The optimal switching index $j^*(k)$ and the optimal feedback matrix $\boldsymbol{K}^*(k)$ are determined at time instant t_k by solving online the LMI optimization problem for the current state $\boldsymbol{x}(k)$. In contrast to the PCS/RHCS strategy where the feedback matrices $\boldsymbol{K}(k)$ are designed offline without considering $\boldsymbol{x}(k)$, they are designed here online while taking the current state $\boldsymbol{x}(k)$ explicitly into account. Since the proposed codesign problem is solved online, its feasibility is further analyzed and a certificate for closed-loop stability is also given. An offline version of the strategy is finally proposed, reducing the computational complexity at the expense of introducing further suboptimality.

Event-Based Control and Scheduling. The codesign strategies outlined above are characterized by requiring state measurements from all plants at each time instant t_k for scheduling purposes. This requirement might not be applicable in some real-world applications, on the one hand, and might even induce a large scheduling overhead which in turn degrade the resulting performance, on the other hand. Alternatively, each plant is assigned an event generator implementing a threshold-based event-triggering law $\sigma_i(k)$, $\forall i \in \mathbb{M}$, for the decision whether to send the current state measurement. The dynamic threshold of each event-triggering law can further be tuned by a design parameter $\lambda_i \in \mathbb{R}_0^+$. The scheduling decision is made based on the output (and not based on the state measurements) of the associated event triggering laws. By introducing event-triggering laws, the main codesign problem is transformed into the problem of determining $\sigma_i(k)$ and the feedback matrices $\boldsymbol{K}_i$ for all $i \in \mathbb{M}$. This problem is formulated as an LMI optimization problem based on the S-procedure. The effectiveness of the proposed strategy and the effect of the chosen tuning parameters λ_i, $\forall i \in \mathbb{M}$, on the resulting control performance are evaluated by simulation for an illustrative example and experimentally for a case study.

Prediction-Based Control and Scheduling. Under all codesign methods outlined above, only one plant can be served at each time instant t_k. This is due to the restriction on the number of available communication channels. Within this strategy, all plants can simultaneously be controlled despite the existence of such restriction. To this end, all control tasks are implemented on one (rather than M) embedded processor. Moreover, a model-based predictor is also implemented on the embedded processor for the purpose of computing the control signals. A replica of the predicted states in the embedded processor is also generated at the associated even generators for evaluation purposes of the state error (difference between actual state and predicted state). If the state error exceeds a specific threshold, the actual state measurement of the most demanding plant is sent first to the embedded processor for correcting the predicted states. Simulation results have shown the advantage of introducing model-based predictors at the plant/controller side in terms of improving the control performance. Experimental results, on the other hand, have shown the disadvantage of considering model-based predictors for control and scheduling if accurate models on system dynamics are not available. In this case, the control signals are computed based on a "wrong" prediction of system states and events are triggered more than required.

Final Word. To sum up, five codesign methods are proposed in this thesis. The first three methods perform scheduling in a centralized fashion while the last two methods in a "decentralized" fashion. Thus, the selection of a specific method depends strongly on the application at hand. For applications where a centralized scheduling is implementable with a small scheduling overhead, the RHCS strategy (or even the IACS strategy if high-performance embedded processors are available) can be used. Otherwise, the EBCS strategy (or even the PBCS strategy if accurate models on system dynamics are available) can be used. All codesign strategies are evaluated and implemented for a case study. Although simulation/experimental results are not evidence, they can give a very good indication of the properties a certain strategy may have. It has been shown that the proposed strategies outperform the well-known TDMA strategy and other strategies proposed in literature such as the OPP strategy. Analytical assessment of the amount of suboptimality induced by the proposed strategies due to the upper bound consideration on the worst-case cost value is one of the interesting topics to be addressed in future work besides other topics that are mentioned in the following section.

10.2 Future Work

Although several methods and concepts have been proposed and studied in this thesis, there are still possible extensions to be explored in future work. Based on the work proposed in this thesis, some suggestions for future work are listed below.

Suboptimality Analysis. Five suboptimal codesign methods are proposed in this thesis to tackle Problem 3.1. Quantifying the amount of suboptimality induced by each of them is of special importance. It serves, on the one hand, for a quantitative assessment of the achievable performance and it can, on the other hand, be used for developing criteria of determining suitable periods (for the PCS strategy), prediction horizons (for the RHCS strategy), and tuning parameters (for the EBCS and PBCS strategy). This issue has been addressed in literature yet for receding-horizon control of linear systems [NP97, PN00, BF03, AHT12, AH14] and nonlinear systems [DM04, GR08, Grü09]. Suboptimality analysis for hybrid systems, mainly switched systems, is still open.

Output-Based Codesign. All codesign methods proposed in this thesis are based on the assumption that all system states are measurable. This assumption of full state feedback information is, however, not always satisfied as the measurement of some states may not be possible or very noisy. Extending the proposed control and scheduling codesign strategies toward output-based rather than state-based is thus certainly desirable. For the IACS strategy, an output-based codesign version has been recently addressed in [RAL13b]. The IACS codesign problem is decomposed into a state-based control and scheduling codesign subproblem and an observer design subproblem using the separation principle. Both subproblems are eventually formulated as LMI subproblems which can be solved efficiently. Two types of observers have been studied to cope with the lack of all states, namely prediction observer and current-state observer. The other codesign strategies can be revised for output-based extension taking the work above and further in [LL11b, DH12, BDWH12, GA12, YA13] as a basis.

Dynamic-Based Codesign. Instead of static state feedback control laws (P-type controllers), control laws with internal dynamics such as PI-type, PID-type, or other dynamic controllers can be considered for more degrees of freedom in parametrizing the controller and thus in stabilizing the controlled system. Within the event-based control framework, an event-triggered PI control design strategy with guaranteed performance is proposed in [RVAL15]. Instead of letting the integral state of the PI controller be updated in a continuous-time manner or only if an event exists as supposed in literature, it is updated at every sampling instant (discrete-time). The update algorithm is executed at the sensor side next to the proposed event generator and sent with new measurements to the PI controller if a necessity exists. The event-triggered PI control design problem is introduced and formulated as an LMI optimization problem. The same way of thinking can be applied to scheduling mechanisms in which switching laws with internal dynamics rather than static stat-based switching laws are to be considered. Particularly, in association with dynamic-based event generators. To this end, the recent work proposed in [DBH14, Gir15] can serve as a good basis.

Constrained Codesign. Control and scheduling codesign subject to input and state constraints that are induced by hardware such as actuator limits or by operational environment such as safety limits is an important direction of future research. Several constrained control design strategies have been proposed in literature for linear systems with finite time-horizon quadratic cost functions [MRRS00, BBM03, PSM04, PSM05, BHA14, BHA15], linear systems with infinite time-horizon quadratic cost functions [KBM96, SR98, WK03, GBTM04], piecewise affine (PWA) hybrid systems [MR03, BCM06], and mixed logical dynamical systems [BM99, LHWB06]. These papers might be a firm basis for extending the codesign strategies proposed in the thesis with respect to input and state constraints.

For the PBCS strategy, an extension toward constrained control and scheduling codesign has been recently studied in [AGL15b]. The constraints are given in the form of linear inequalities that constitute polyhedral sets containing the origin. The codesign problem subject to the input and state linear inequality constraints is formulated as an LMI optimization problem with the objective of maximizing the domain of attraction (an invariant subset of the state space in which the input and state constraints are satisfied). Similar ideas might be examined for the other codesign strategies proposed in the thesis.

Stochastic Codesign. In networked control systems, almost all network-induced imperfections can be described by stochastic processes such as Bernoulli process or Markov process. For instance, variable transmission times and variable sampling instants have been represented as stochastic variables characterized by Markov chains in [MA03, ZSCH05, SY09, LZJ09] or by stationary processes in [NBW98, DHB+12]. Similar stochastic representations have been proposed for packet dropouts in [WC07, SSF+07], quantization errors in [NE03, NJWY09], and medium access constraints in [TN08, CH08, MH11]. Adopting these representations within the control and scheduling codesign framework is an interesting direction of future work that requires fundamental research.

Recently, a stochastic event-based control and scheduling codesign strategy for networked control systems with variable transmission times has been proposed in [AGL14, AGL15a]. The variable transmission time is uniquely represented as an uncertain time-varying input delay that belongs to a finite set of different bounded intervals. The transition among the bounded intervals is described by a stochastic process. Two stochastic processes have been considered in this work, namely a Markov process and a stationary process. Mean-square stability with guaranteed performance (measured by a quadratic cost function) is provided for both versions after formulating the codesign problem as an LMI optimization problem. Although the work is very promising, further research for addressing the other network-induced imperfections from stochastic perspectives is necessary.

Distributed Codesign. The case of dynamically decoupled subsystems is addressed in this thesis. In case of dynamically coupled subsystems, distributed control and scheduling codesign becomes a more important concept to be addressed from both stability and performance perspectives. To this end, the proposed modeling and codesign strategies must be extended. While extensions of the modeling procedure for dynamically coupled subsystems are quite straightforward, see [AGL13b, AGL15c], extensions of the proposed centralized/decentralized codesign strategies toward a distributed architecture require fundamental research. A good basis for this direction of future research might be the work proposed in [MR09, DMUH14] for time-triggered distributed control and in [WL11, GDJ+13, SL13] for event-triggered distributed control.

A Appendix

A.1 Matrix Polynomial Overapproximation

Lemma A.1 ([HDL07]) *Consider a matrix polynomial*

$$\mathbf{\Gamma}(\rho) = \sum_{\ell=1}^{L} \rho^\ell \mathbf{\Gamma}_\ell \tag{A.1}$$

with the uncertain parameter $\rho \in [\underline{\rho}, \overline{\rho}]$, $\underline{\rho} \in \mathbb{R}_0^+$, *and the constant matrices* $\mathbf{\Gamma}_\ell$. *There exists then a convex polytope with* $L+1$ *vertices*

$$\begin{aligned}
\hat{\mathbf{\Gamma}}_1 &= \underline{\rho}^L \mathbf{\Gamma}_L + \underline{\rho}^{L-1} \mathbf{\Gamma}_{L-1} + \ldots + \underline{\rho}^2 \mathbf{\Gamma}_2 + \underline{\rho} \mathbf{\Gamma}_1 \\
\hat{\mathbf{\Gamma}}_2 &= \underline{\rho}^L \mathbf{\Gamma}_L + \underline{\rho}^{L-1} \mathbf{\Gamma}_{L-1} + \ldots + \underline{\rho}^2 \mathbf{\Gamma}_2 + \overline{\rho} \mathbf{\Gamma}_1 \\
\hat{\mathbf{\Gamma}}_3 &= \underline{\rho}^L \mathbf{\Gamma}_L + \underline{\rho}^{L-1} \mathbf{\Gamma}_{L-1} + \ldots + \overline{\rho}^2 \mathbf{\Gamma}_2 + \overline{\rho} \mathbf{\Gamma}_1 \\
&\vdots \\
\hat{\mathbf{\Gamma}}_{L+1} &= \overline{\rho}^L \mathbf{\Gamma}_L + \overline{\rho}^{L-1} \mathbf{\Gamma}_{L-1} + \ldots + \overline{\rho}^2 \mathbf{\Gamma}_2 + \overline{\rho} \mathbf{\Gamma}_1
\end{aligned} \tag{A.2}$$

that envelopes the matrix polynomial $\mathbf{\Gamma}(\rho)$, *i.e.*

$$\mathbf{\Gamma}(\rho) = \sum_{\ell=1}^{L+1} \mu_\ell(\rho) \hat{\mathbf{\Gamma}}_\ell \quad \text{where} \quad \sum_{\ell=1}^{L+1} \mu_\ell(\rho) = 1 \quad \text{and} \quad \mu_\ell(\rho) \geq 0 \quad \forall \ell. \tag{A.3}$$

Moreover, the uncertain parameters $\mu_\ell(\rho)$ *depend only on* ρ *and are independent of* $\mathbf{\Gamma}_\ell$.

PROOF. Substituting (A.1) into (A.3) yields

$$\sum_{\ell=1}^{L} \rho^\ell \mathbf{\Gamma}_\ell = \sum_{\ell=1}^{L+1} \mu_\ell(\rho) \hat{\mathbf{\Gamma}}_\ell. \tag{A.4}$$

Substituting further (A.2) into (A.4) while taking the restriction $\sum_{\ell=1}^{L+1} \mu_\ell(\rho) = 1$ into account leads to

$$\begin{pmatrix} 1 & \cdots & \cdots & \cdots & 1 \\ \underline{\rho} & \overline{\rho} & \cdots & \cdots & \overline{\rho} \\ \underline{\rho}^2 & \underline{\rho}^2 & \overline{\rho}^2 & \cdots & \overline{\rho}^2 \\ \vdots & & & \ddots & \vdots \\ \underline{\rho}^L & \cdots & \cdots & \underline{\rho}^L & \overline{\rho}^L \end{pmatrix} \begin{pmatrix} \mu_1 \\ \mu_2 \\ \vdots \\ \mu_{L+1} \end{pmatrix} = \begin{pmatrix} 1 \\ \rho \\ \rho^2 \\ \vdots \\ \rho^L \end{pmatrix}. \tag{A.5}$$

What left is just to prove that there exists a set of non-negative uncertain parameters $\mu_\ell \in \mathbb{R}_0^+$, $\ell \in \{1, \dots, L+1\}$, which satisfies the linear system (A.5). Since the columns of the system matrix are linearly independent, its determinant is not null and thus a unique solution $\begin{pmatrix}\mu_1 & \cdots & \mu_{L+1}\end{pmatrix}^T$ exists. Using the classical Gaussian elimination method [Mey00], the solution of (A.5) is given by

$$\begin{aligned} \mu_1 &= 1 - \frac{\rho - \underline{\rho}}{\overline{\rho} - \underline{\rho}} \\ \mu_\ell &= \frac{\rho^{\ell-1} - \underline{\rho}^{\ell-1}}{\overline{\rho}^{\ell-1} - \underline{\rho}^{\ell-1}} - \frac{\rho^{\ell} - \underline{\rho}^{\ell}}{\overline{\rho}^{\ell} - \underline{\rho}^{\ell}} \qquad \forall \ell \in \{2, \dots, L+1\}. \end{aligned} \tag{A.6}$$

It can be noted that the solution given in (A.6) depends only on the uncertain parameter $\rho \in [\underline{\rho}, \overline{\rho}]$. Moreover, all elements μ_ℓ, $\ell \in \{1, \dots, L+1\}$, are non-negative due to the fact that the function $f : \mathbb{R}_0^+ \to \mathbb{R}_0^+$, $f(x) = \frac{\rho^x - \underline{\rho}^x}{\overline{\rho}^x - \underline{\rho}^x}$, is non-negative and monotonically decreasing for $x \in \mathbb{R}^+$. This completes the proof. □

A.2 Robust LMI for Bounded Uncertainty

Lemma A.2 ([Löf03]) *Let $\boldsymbol{Y}$, $\boldsymbol{M}$ and $\boldsymbol{N}$ be given constant matrices of compatible dimension, then for any uncertain matrix $\boldsymbol{E}$ satisfying $\|\boldsymbol{E}\|_2 \leq 1$,*

$$\boldsymbol{Y} + \boldsymbol{M}\boldsymbol{E}\boldsymbol{N} + \boldsymbol{N}^T\boldsymbol{E}^T\boldsymbol{M}^T \succeq \boldsymbol{0} \tag{A.7}$$

is equivalent to

$$\boldsymbol{Y} - \beta\boldsymbol{M}\boldsymbol{M}^T - \beta^{-1}\boldsymbol{N}^T\boldsymbol{N} \succeq \boldsymbol{0} \tag{A.8}$$

where $\beta \in \mathbb{R}_0^+$ is a non-negative real scalar.

Proof. Inequality (A.7) is equivalent to

$$\boldsymbol{x}^T(\boldsymbol{Y} + \boldsymbol{M}\boldsymbol{E}\boldsymbol{N} + \boldsymbol{N}^T\boldsymbol{E}^T\boldsymbol{M}^T)\boldsymbol{x} \geq 0 \qquad \forall \boldsymbol{x}. \tag{A.9}$$

Introducing the vector $\boldsymbol{y} = \boldsymbol{E}^T\boldsymbol{M}^T\boldsymbol{x}$, (A.9) can be rewritten as

$$\boldsymbol{x}^T\boldsymbol{Y}\boldsymbol{x} + \boldsymbol{y}^T\boldsymbol{N}\boldsymbol{x} + \boldsymbol{x}^T\boldsymbol{N}^T\boldsymbol{y} \geq 0. \tag{A.10}$$

Inequality (A.10) must hold whenever $\|\boldsymbol{E}\|_2 \leq 1$, i.e. whenever $\boldsymbol{E}\boldsymbol{E}^T \preceq \boldsymbol{I}$. In terms of the vectors $\boldsymbol{x}$ and $\boldsymbol{y}$, this implies that (A.10) must hold whenever

$$\boldsymbol{y}^T\boldsymbol{y} \leq \boldsymbol{x}^T\boldsymbol{M}\boldsymbol{M}^T\boldsymbol{x}. \tag{A.11}$$

Thus, the satisfaction of (A.7) whenever $\|\boldsymbol{E}\|_2 \leq 1$ is equivalent to the satisfaction of (A.10) whenever (A.11) holds, i.e.

$$\begin{pmatrix}\boldsymbol{x}\\ \boldsymbol{y}\end{pmatrix}^T \begin{pmatrix}\boldsymbol{Y} & \boldsymbol{N}^T\\ \boldsymbol{N} & \boldsymbol{0}\end{pmatrix}\begin{pmatrix}\boldsymbol{x}\\ \boldsymbol{y}\end{pmatrix} \geq 0 \quad \text{whenever} \quad \begin{pmatrix}\boldsymbol{x}\\ \boldsymbol{y}\end{pmatrix}^T \begin{pmatrix}\boldsymbol{M}\boldsymbol{M}^T & \boldsymbol{0}\\ \boldsymbol{0} & -\boldsymbol{I}\end{pmatrix}\begin{pmatrix}\boldsymbol{x}\\ \boldsymbol{y}\end{pmatrix} \geq 0. \tag{A.12}$$

Applying the S-procedure, (A.12) can equivalently be rewritten as

$$\begin{pmatrix}\boldsymbol{x}\\ \boldsymbol{y}\end{pmatrix}^T \begin{pmatrix}\boldsymbol{Y} - \beta\boldsymbol{M}\boldsymbol{M}^T & \boldsymbol{N}^T\\ \boldsymbol{N} & \beta\boldsymbol{I}\end{pmatrix}\begin{pmatrix}\boldsymbol{x}\\ \boldsymbol{y}\end{pmatrix} \geq 0. \tag{A.13}$$

where $\beta \in \mathbb{R}_0^+$ is a real scalar to be determined. Applying further the Schur complement to (A.13) leads finally to (A.8). This completes the proof. □

A.3 The S-Procedure Relaxation

The S-procedure is nothing more than a Lagrange relaxation technique that is mostly used in problems with quadratic constraints [JÖ6]. It enables one to combine several quadratic inequalities into one single inequality, with some conservatism in general. It is frequently used in system theory to derive sufficient stability conditions in terms of LMIs for a large number of nonconvex problems. To illustrate the basic idea behind the S-procedure, let's consider a quadratic objective function of $\boldsymbol{x} \in \mathbb{R}^n$

$$F_0(\boldsymbol{x}) \triangleq \boldsymbol{x}^T\boldsymbol{A}_0\boldsymbol{x} + 2\boldsymbol{b}_0^T\boldsymbol{x} + c_0 \leq 0 \tag{A.14}$$

subject to quadratic constraints

$$F_i(\boldsymbol{x}) \triangleq \boldsymbol{x}^T\boldsymbol{A}_i\boldsymbol{x} + 2\boldsymbol{b}_i^T\boldsymbol{x} + c_i \geq 0 \qquad \forall i \in \{1, 2, \dots, q\}. \tag{A.15}$$

The quadratic functions $F_i(\boldsymbol{x})$, $\forall i \in \{0, \dots, q\}$, are not convex functions in general, making the problem above an NP-hard problem. A more conservative but simpler to solve setup is to consider an augmented (unconstrained) quadratic objective

$$F_{\text{aug}}(\boldsymbol{x}) \triangleq F_0(\boldsymbol{x}) + \sum_{i=1}^{q} \beta_i F_i(\boldsymbol{x}) \leq 0 \qquad \forall \boldsymbol{x} \in \mathbb{R}^n \tag{A.16}$$

where $\beta_i \in \mathbb{R}_0^+$, $\forall i \in \{1, \dots, q\}$, are non-negative scalars to be determined. Obviously, the satisfaction of (A.16) implies the satisfaction of (A.14) subject to (A.15). The condition (A.16) is generally a sufficient condition and not necessary. Exception is the case when $q = 1$ in which (A.16) becomes necessary and sufficient.

Note that the quadratic objective (A.16) can equivalently be cast as a convex LMI problem

$$\begin{pmatrix} \boldsymbol{A}_0 & \boldsymbol{b}_0 \\ \boldsymbol{b}_0^T & c_0 \end{pmatrix} + \sum_{i=1}^{q} \beta_i \begin{pmatrix} \boldsymbol{A}_i & \boldsymbol{b}_i \\ \boldsymbol{b}_i^T & c_i \end{pmatrix} \preceq \boldsymbol{0}, \qquad \beta_i \geq 0. \tag{A.17}$$

Further details on LMIs and related topics such as the most commonly used LMI tricks (e.g. the congruence transformation or the Schur complement) are reported in [SP05, Chapter 12] and in [BGFB94].

Bibliography

[AA14] Erling D. Andersen and Knud D. Andersen. MOSEK: Matlab optimization toolbox, version 7. http://www.mosek.com, 2014.

[AB04] Panos Antsaklis and John Baillieul. Special issue on networked control systems. *IEEE Transactions on Automatic Control*, 49(9):1421–1423, 2004.

[AB07] Panos Antsaklis and John Baillieul. Special issue on technology of networked control systems. *Proceedings of the IEEE*, 95(1):5–8, 2007.

[ÅC05] Karl-Erik Årzén and A. Cervin. Control and embedded computing: Survey of research directions. In *Proceedings of the 16th IFAC World Congress*, 2005.

[ÅCES00] Karl-Erik Årzén, Anton Cervin, Johan Eker, and Lui Sha. An introduction to control and scheduling co-design. In *Proceedings of the 39th IEEE Conference on Decision and Control*, pages 4865–4870, 2000.

[AGL11] Sanad Al-Areqi, Daniel Görges, and Steven Liu. Robust control and scheduling codesign for networked embedded control systems. In *Proceedings of the 50th IEEE Conference on Decision and Control and European Control Conference*, pages 3154–3159, 2011.

[AGL12a] Sanad Al-Areqi, Daniel Görges, and Steven Liu. Receding-horizon control and scheduling of systems with uncertain computation and communication delays. In *Proceedings of the 51st IEEE Conference on Decision and Control*, pages 2654–2659, 2012.

[AGL12b] Sanad Al-Areqi, Daniel Görges, and Steven Liu. Robust feedback control and scheduling of networked embedded control systems. In *Proceedings of the 4th IFAC Conference on Analysis and Design of Hybrid Systems*, pages 127–132, 2012.

[AGL13a] Sanad Al-Areqi, Daniel Görges, and Steven Liu. Event-based control and scheduling codesign of networked embedded control systems. In *Proceedings of the American Control Conference*, pages 52999–5304, 2013.

[AGL13b] Sanad Al-Areqi, Daniel Görges, and Steven Liu. Robust event-based control and scheduling of networked embedded control systems. In *Proceedings of the 4th IFAC Workshop on Distributed Estimation and Control in Networked Systems*, pages 7–14, 2013.

[AGL14] Sanad Al-Areqi, Daniel Görges, and Steven Liu. Stochastic event-based control and scheduling of large-scale networked control systems. In *Proceedings of the European Control Conference*, pages 2316–2321, 2014.

[AGL15a] Sanad Al-Areqi, Daniel Görges, and Steven Liu. Event-based control and scheduling codesign: Stochastic and robust approaches. *IEEE Transactions on Automatic Control*, 60(5):1291–1303, 2015.

[AGL15b] Sanad Al-Areqi, Daniel Görges, and Steven Liu. Event-based control and scheduling codesign subject to input and state constraints. In *Proceedings of the 2015 European Control Conference*, pages 1860–1865, 2015.

[AGL15c] Sanad Al-Areqi, Daniel Görges, and Steven Liu. Event-based networked control and scheduling codesign with guaranteed performance. *Automatica*, 57:128–134, 2015.

[AGM07] Jeffrey G. Andrews, Arunabha Ghosh, and Rias Muhamed. *Fundamentals of WiMAX: Understanding Broadband Wireless Networking*. Pearson Education, 2007.

[AH14] Duarte Antunes and W. P. M. H. Heemels. Rollout event-triggered control: Beyond periodic control performance. *IEEE Transactions on Automatic Control*, 59(12):3296–3311, 2014.

[AHT12] Duarte Antunes, W. P. M. H. Heemels, and P. Tabuada. Dynamic programming formulation of periodic event-triggered control: performance guarantees and co-design. In *Proceedings of the 51st IEEE Conference on Decision and Control*, pages 7212–7217, 2012.

[Årz99] Karl-Erik Årzén. A simple event-based PID controller. In *Proceedings of the 14th IFAC World Congress*, pages 423–428, 1999.

[Åst08] Karl J. Åström. Event based control. In *Analysis and Design of Nonlinear Control Systems*, pages 127–147. Springer, 2008.

[ÅW97] Karl J. Åström and Björn Wittenmark. *Computer-Controlled Systems: Theory and Design*. Prentice-Hall, Upper Saddle River, NJ, 3rd edition, 1997.

[BA02] Alex Brand and Hamid Aghvami. *Multiple access protocols for mobile communications: GPRS, UMTS and beyond*. John Wiley & Sons, England, 2002.

[BA11] Rainer Blind and Frank Allgöwer. Analysis of networked event-based control with a shared communication medium: Part I – Pure ALOHA. In *Proceedings of the 18th IFAC World Congress*, 2011.

[BA12a] Rainer Blind and Frank Allgöwer. Is it worth to retransmit lost packets in networked control systems? In *Proceedings of the 51st IEEE Conference on Decision and Control*, pages 1368–1373, 2012.

[BA12b] Rainer Blind and Frank Allgöwer. The performance of event-based control for scalar systems with packet losses. In *Proceedings of the 51st IEEE Conference on Decision and Control*, pages 6572–6576, 2012.

[BB09] Davide Barcelli and Alberto Bemporad. Decentralized model predictive control of dynamically-coupled linear systems: Tracking under packet loss. In *Proceedings of the 1st IFAC Workshop on Estimation and Control of Networked Systems*, pages 204–209, 2009.

[BBM03] Alberto Bemporad, Francesco Borrelli, and Manfred Morari. Min-max control of constrained uncertain discrete-time linear systems. *IEEE Transactions on Automatic Control*, 48(9):1600–1606, 2003.

[BÇH06] Mohamed El Mongi Ben Gaid, Arben Çela, and Yskandar Hamam. Optimal integrated control and scheduling of networked control systems with communication constraints: Application to a car suspension system. *IEEE Transactions on Control Systems Technology*, 14(4):776–787, 2006.

[BÇH09] Mohamed El Mongi Ben Gaid, Arben Çela, and Yskandar Hamam. Optimal real-time scheduling of control tasks with state feedback resource allocation. *IEEE Transactions on Control Systems Technology*, 17(2):309–326, 2009.

[BCM06] Mato Baotić, Frank J. Christophersen, and Manfred Morari. Constrained optimal control of hybrid systems with a linear performance index. *IEEE Transactions on Automatic Control*, 51(12):1903–1919, 2006.

[BDWH12] N. W. Bauer, M. C. F. Donkers, N. van de Wouw, and W. P. M. H. Heemels. Decentralized static output-feedback control via networked communication. In *Proceedings of the American Control Conference*, pages 5700–5705, 2012.

[Bel57] Richard E. Bellman. *Dynamic Programming*. Princeton University Press, Princeton, NJ, 1957.

[Ber05] Dimitri P. Bertsekas. *Dynamic Programming and Optimal Control*, volume 1. Athena Scientific, Belmont, MA, 3rd edition, 2005.

[BF03] Alberto Bemporad and Carlo Filippi. Suboptimal explicit receding horizon control via approximate multiparametric quadratic programming. *Journal of Optimization Theory and Applications*, 117(1):9–38, 2003.

[BGFB94] Stephen Boyd, Laurent El Ghaoui, Eric Feron, and Venkataramanan Balakrishnan. *Linear Matrix Inequalities in System and Control Theory*. Society for Industrial and Applied Mathematics, 1994.

[BHA14] Florian D. Brunner, W. P. M. H. Heemels, and Frank Allgöwer. Robust self-triggered mpc for constrained linear systems. In *Proceeding of the European Control Conference*, pages 472–477, 2014.

[BHA15] Florian D. Brunner, W. P. M. H. Heemels, and Frank Allgöwer. Robust event-triggered mpc for constrained linear discrete-time systems with guaranteed average sampling rate. In *Proceeding of the IFAC Conference on Nonlinear Model Predictive Control*, 2015.

[BHJ10a] A. Bemporad, M. Heemels, and M. Johansson. *Networked Control Systems*, chapter 7: Stability and Stabilization of Networked Control Systems, pages 203–253. Springer-Verlag Berlin Heidelberg, 2010.

[BHJ10b] Alberto Bemporad, Maurice Heemels, and Mikael Johansson. *Networked Control Systems*. Springer-Verlag Berlin Heidelberg, 2010.

[BK08] Ajinkya Y. Bhave and Bruce H. Krogh. Performance bounds on state-feedback controllers with network delay. In *Proceedings of the 47th IEEE Conference on Decision and Control*, pages 4608–4613, 2008.

[BL00] Roger W. Brockett and Daniel Liberzon. Quantized feedback stabilization of linear systems. *IEEE Transactions on Automatic Control*, 45(7):1279–1289, 2000.

[BLS07] Martin Buehler, Karl Lagnemma, and Sanjiv Singh, editors. *The 2005 DARPA Grand Challenge: The great robot race*. Springer Science & Business Media, 2007.

[BM99] Alberto Bemporad and Manfred Morari. Control of systems integrating logic, dynamics, and constraints. *Automatica*, 35(3):407–427, 1999.

[BPZ02] Michael S. Branicky, Stephen M. Phillips, and Wei Zhang. Scheduling and feedback co-design for networked control systems. In *In Proceedings of 41st IEEE Conference on Decision and Control*, pages 1211–1217, 2002.

[Bra98] Michael S. Branicky. Multiple Lyapunov functions and other analysis tools for switched and hybrid systems. *IEEE Transactions on Automatic Control*, 43(4):475–482, 1998.

[BRRA09] Christoph Böhm, Tobias Raff, Marcus Reble, and Frank Allgöwer. LMI-based model predictive control for linear discrete-time periodic systems. In *Nonlinear Model Predictive Control*, pages 99–108, 2009.

[Bus01] Linda G. Bushnell. Networks and control. *IEEE Control Systems Magazine*, 21(1):22–23, 2001.

[CA06] Anton Cervin and Peter Alriksson. Optimal on-line scheduling of multiple control tasks: A case study. In *Proceedings of the 18th Euromicro Conference on Real-Time Systems*, pages 141–150, 2006.

[CGA08] Patrizio Colaneria, José C. Geromel, and Alessandro Astolfi. Stabilization of continuous-time switched nonlinear systems. *Systems & Control Letters*, 57(1):95–103, 2008.

[CH08] Anton Cervin and Toivo Henningsson. Scheduling of event-triggered controllers on a shared network. In *Proceedings of the 47th IEEE Conference on Decision and Control*, pages 360–366, 2008.

[CHL+03] Anton Cervin, Dan Henriksson, Bo Lincoln, Johan Eker, and Karl-Erik Årzén. How does control timing affect performance? *IEEE Control Systems Magazine*, 23(3):16–30, 2003.

[CJ08] Anton Cervin and Erik Johannesson. Sporadic control of scalar systems with delay, jitter and measurement noise. In *Proceedings of the 17th IFAC World Congress*, pages 15486–15492. 2008.

[CWHN06] Marieke B.G. Cloosterman, Nathan van de Wouw, W. P. M. H. Heemels, and H. Nijmeijer. Robust stability of networked control systems with time-varying network-induced delays. In *Proceedings of the 45th IEEE Conference on Decision and Control*, pages 4980–4985, San Diego, CA, USA, 2006.

[CWHN09] Marieke B. G. Cloosterman, Nathan van de Wouw, W. P. M. H. Heemels, and Hendrik Nijmeijer. Stability of networked control systems with uncertain time-varying delays. *IEEE Transactions on Automatic Control*, 54(7):1575–1580, 2009.

[D'A05] Raffaello D'Andrea. *The Cornell RoboCup Robot Soccer Team: 1999-2003*, chapter 6, pages 793–804 Birkhäuser Boston, 2005.

[DB01] Jamal Daafouz and Jacques Bernussou. Parameter dependent Lyapunov functions for discrete time systems with time varying parametric uncertainties. *Systems & Control Letters*, 43(5):355–359, 2001.

[DBH14] Victor S. Dolk, D. P. Borgers, and W. P. M. H. Heemels. Dynamic event-triggered control: Tradeoffs between transmission intervals and performance. In *Proceedings of the 53rd IEEE Conference on Decision and Control*, pages 2764–2769, 2014.

[DBPL00] Raymond A. DeCarlo, Michael S. Branicky, Stefan Pettersson, and Bengt Lennartson. Perspectives and results on the stability and stabilizability of hybrid systems. *Proceedings of the IEEE*, 88(7):1069–1082, 2000.

[DFH13] Maxim Dolgov, Jörg Fischer, and Uwe D. Hanebeck. Event-based LQG control over networks with random transmission delays and packet losses. In *Proceedings of the 4th IFAC Workshop on Distributed Estimation and Control in Networked Systems*, pages 23–30, 2013.

[DH12] M. C. F. Donkers and W. P. M. H. Heemels. Output-based event-triggered control with guaranteed $\mathcal{L}_\infty$-Gain and improved and decentralized event-triggering. *IEEE Transactions on Automatic Control*, 57(6):1362–1376, 2012.

[DHB+12] M. C. F. Donkers, W. P. M. H. Heemels, D. Bernardini, A. Bemporad, and V. Shneer. Stability analysis of stochastic networked control systems. *Automatica*, 48(5):917–925, 2012.

[DHWH11] M. C. F. Donkers, W. P. M. H. Heemels, Nathan van de Wouw, and Laurentiu Hetel. Stability analysis of networked control systems using a switched linear systems approach. *IEEE Transactions on Automatic Control*, 56(9):2101–2115, 2011.

[DLG10] Shi-Lu Dai, Hai Lin, and Shuzhi Sam Ge. Scheduling-and-control codesign for a collection of networked control systems with uncertain delays. *IEEE Transactions on Control Systems Technology*, 18(1):66–78, 2010.

[DM04] Federico Di Palma and Lalo Magni. On optimality of nonlinear model predictive control. In *Proceedings of the 16th International Symposium on Mathematical Theory of Networks and Systems*, 2004. See also Systems & Control Letters, Volume 56, Issue 1, January 2007, Pages 58–61.

[DMUH14] Frederik Deroo, Martin Meinel, Michael Ulbrich, and Sandra Hirche. Distributed control design with local model information and guaranteed stability. In *Proceedings of the 19th IFAC World Congress*, pages 4010–4017, 2014.

[DN07] Dragan B. Dačić and Dragan Nešić. Quadratic stabilization of linear networked control systems via simultaneous protocol and controller design. *Automatica*, 43(7):1145–1155, 2007.

[DRL02] Jamal Daafouz, Pierre Riedinger, and Claude Lung. Stability analysis and control synthesis for switched systems: A switched Lyapunov function approach. *IEEE Transactions on Automatic Control*, 47(11):1883–1887, 2002.

[DT00] Carlos E. De Souza and Alexandre Trofino. An LMI approach to stabilization of linear discrete-time periodic systems. *International Journal of Control*, 73(8):696–703, 2000.

[Dur13] Sylvain Durand. Event-based stabilization of linear system with communication delays in the measurements. In *Proceedings of the American Control Conference*, pages 152–157, 2013.

[FD09] Emilia Fridman and Michel Dambrine. Control under quantization, saturation and delay: An LMI approach. *Automatica*, 45(10):2258–2264, 2009.

[FHDH13] Jörg Fischer, Achim Hekler, Maxim Dolgov, and Uwe D. Hanebeck. Optimal sequence-based LQG control over TCP-like networks subject to random transmission delays and packet losses. In *Proceedings of the American Control Conference*, pages 1546–1552, 2013.

[FMH+14] Filiberto Fele, Jose M. Maestre, S. Mehdy Hashemy, David Mu noz de la

Peña, and Eduardo F. Camacho. Coalitional model predictive control of an irrigation canal. *Journal of Process Control*, 24(4):314–325, 2014.

[FPW90] Gene F. Franklin, J. David Powell, and Michael L. Workman. *Digital control of dynamic systems.* Addison-Wesley, Reading, MA, 2nd edition, 1990.

[FSR04] Emilia Fridman, Alexandre Seuret, and Jean-Pierre Richard. Robust sampled-data stabilization of linear systems: An input delay approach. *Automatica*, 40(8):1441–1446, 2004.

[Fuj09] Hisaya Fujioka. A discrete-time approach to stability analysis of systems with aperiodic sample-and-hold devices. *IEEE Transactions on Automatic Control*, 54(10):2440–2445, 2009.

[GA11] Eloy Garcia and Panos J. Antsaklis. Model-based event-triggered control with time-varying network delays. In *Proceedings of the 50th IEEE Conference on Decision and Control and European Control Conference*, pages 1650–1655, 2011.

[GA12] Eloy Garcia and Panos J. Antsaklis. Output feedback model-based control of uncertain discrete-time systems with network induced delays. In *Proceedings of the 51st IEEE Conference on Decision and Control*, pages 6647–6652, 2012.

[GB08] Michael Grant and Stephen Boyd. Graph implementations for nonsmooth convex programs. In V. Blondel, S. Boyd, and H. Kimura, editors, *Recent Advances in Learning and Control*, Lecture Notes in Control and Information Sciences, pages 95–110. Springer-Verlag Limited, 2008. `http://stanford.edu/~boyd/graph_dcp.html`.

[GB14] Michael Grant and Stephen Boyd. CVX: Matlab software for disciplined convex programming, version 2.1. `http://cvxr.com/cvx`, 2014.

[GBTM04] Pascal Grieder, Francesco Borrelli, Fabio Torrisi, and Manfred Morari. Computation of the constrained infinite time linear quadratic regulator. *Automatica*, 40(4):701–708, 2004.

[GC06a] José C. Geromel and Patrizio Colaneri. Stability and stabilization of continuous-time switched linear systems. *SIAM Journal on Control and Optimization*, 45(5):1915–1930, 2006.

[GC06b] José C Geromel and Patrizio Colaneri. Stability and stabilization of discrete time switched systems. *International Journal of Control*, 79(7):719–728, 2006.

[GC08] Huijun Gao and Tongwen Chen. A new approach to quantized feedback control systems. *Automatica*, 44(2):534–542, 2008.

[GC10] Rachana Ashok Gupta and Mo-Yuen Chow. Networked control system:

Overview and research trends. *IEEE Transaction on Industrial Electronics*, 57(7):2527–2535, 2010.

[GCL08] Huijun Gao, Tongwen Chen, and James Lam. A new delay system approach to network-based control. *Automatica*, 44(1):39–52, 2008.

[GDJ+13] María Guinaldo, Dimos V. Dimarogonas, Karl H. Johansson, José Sánchez, and Sebastián Dormido. Distributed event-based control strategies for interconnected linear systems. *IET Control Theory and Applications*, 7(6):877–886, 2013.

[GHBW12] T. M. P. Gommans, W. P. M. H. Heemels, N. W. Bauer, and N. van de Wouw. Compensation-based control for lossy communication networks. In *Proceedings of the American Control Conference*, pages 2854–2859, 2012.

[GHQW04] Graham C. Goodwin, Hernan Haimovich, Daniel E. Quevedo, and James S. Welsh. A moving horizon approach to networked control system design. *IEEE Transactions on Automatic Control*, 49(9):1427–1445, 2004.

[GIL07] Daniel Görges, Michal Izák, and Steven Liu. Optimal control of systems with resource constraints. In *Proceedings of the 46th IEEE Conference on Decision and Control*, pages 1070–1075, 2007.

[GIL09] Daniel Görges, Michal Izák, and Steven Liu. Optimal control and scheduling of networked control systems. In *Proceedings of the 48th IEEE Conference on Decision and Control and 28th Chinese Control Conference*, pages 5839–5844, 2009.

[GIL11] Daniel Görges, Michal Izák, and Steven Liu. Optimal control and scheduling of switched systems. *IEEE Transactions on Automatic Control*, 56(1):135–140, 2011.

[Gir15] Antoine Girard. Dynamic triggering mechanisms for event-triggered control. *IEEE Transactions on Automatic Control (to appear)*, 2015.

[GLS+12] María Guinaldo, Daniel Lehmann, José Sánchez, Sebastián Dormido, and Karl H. Johansson. Distributed event-triggered control with network delays and packet losses. In *Proceedings of the 51st IEEE Conference on Decision and Control*, pages 1–6, 2012.

[GM09] Lars Grüne and Florian Müller. An algorithm for event-based optimal feedback control. In *Proceedings of the 48th IEEE Conference on Decision and Control*, pages 5311–5316, 2009.

[GOL+10] R.H. Gielen, S. Olaru, M. Lazar, W.P.M.H. Heemels, N. van de Wouw, and S.-I. Niculescu. On polytopic inclusions as a modeling framework for systems with time-varying delays. *Automatica*, 46(3):615–619, 2010.

[GPRP13] Konstantinos Gatsis, Miroslav Pajic, Alejandro Ribeiro, and George J.

Pappes. Power-aware communication for wireless sensor-actuator systems. In *Proceedings of the 52nd IEEE Conference on Decision and Control*, pages 4006–4011, 2013.

[GR08] Lars Grüne and Anders Rantzer. On the infinite horizon performance of receding horizon controllers. *IEEE Transactions on Automatic Control*, 53(9):2100–2111, 2008.

[Grü09] Lars Grüne. Analysis and design of unconstrained nonlinear MPC schemes for finite and infinite dimensional systems. *SIAM Journal on Control and Optimization*, 48(2):1206–1228, 2009.

[Gör12] Daniel Görges. *Optimal Control of Switched Systems with Application to Networked Embedded Control Systems*. PhD thesis, Institute of Control Systems, University of Kaiserslautern, Germany, 2012.

[GRB07] Matías García-Rivera and Antonio Barreiro. Analysis of networked control systems with drops and variable delays. *Automatica*, 43(12):2054–2059, 2007.

[HBSW07] Li-Sheng Hu, Tao Bai, Peng Shi, and Ziming Wu. Sampled-data control of networked linear control systems. *Automatica*, 43(5):903–911, 2007.

[HD13] W. P. M. H. Heemels and M. C. F. Donkers. Model-based periodic event-triggered control for linear systems. *Automatica*, 49(3):698–711, 2013.

[HDL06] Laurentiu Hetel, Jamal Daafouz, and Claude Lung. Stabilization of arbitrary switched linear systems with unknown time-varying delays. *IEEE Transactions on Automatic Control*, 51(10):1668–1674, 2006.

[HDL07] Laurentiu Hetel, Jamal Daafouz, and Claude Lung. LMI control design for a class of exponential uncertain systems with application to network controlled switched systems. In *Proceedings of the 2007 American Control Conference*, pages 1401–1406, 2007.

[HDT12] W. P. M. H. Heemels, M. C. F. Donkers, and A. R. Teel. Periodic event-triggered control for linear systems. *IEEE Transactions on Automatic Control*, 58(4):847–861, 2012.

[Het07] Laurentiu Hetel. *Robust stability and control of switched linear systems*. PhD thesis, Graduate School IAEM Lorraine, Nancy University, 2007.

[HJT12] W. P. M. H. Heemels, K. H. Johansson, and P. Tabuada. An introduction to event-triggered and self-triggered control. In *Proceedings of the 51st IEEE Conference on Decision and Control*, pages 3270–3285, 2012.

[HNTW09] W. P. M. H. Heemels, D. Nešić, A. R. Teel, and N. van de Wouw. Networked and quantized control systems with communication delays. In *Proceedings of the 48th IEEE Conference on Decision and Control and 28th Chinese Control Conference*, pages 7929–7935, 2009.

[HNX07] João P. Hespanha, Payam Naghshtabrizi, and Yonggang Xu. A survey of recent results in networked control systems. *Proceedings of the IEEE*, 95(1):138–162, 2007.

[HOV02] João Hespanha, Antonio Ortega, and Lavanya Vasudevan. Towards the control of linear systems with minimum bit-rate. In *Proceedings of the 15th Int. Symp. Mathematical Theory of Networks and Systems*, 2002.

[HQSJ12] Erik Henriksson, Daniel E. Quevedo, Henrik Sandberg, and Karl Henrik Johansson. Self-triggered model predictive control for network scheduling and control. In *Proceedings of the 8th IFAC on Advanced Control of Chemical Processes*, pages 432–438, 2012.

[HSJW03] W. P. M. H. Heemels, H. B. Siahaan, A. Lj. Juloski, and S. Weiland. Control of quantized linear systems: An l_1-optimal control approach. In *Proceedings of the American Control Conference*, pages 3502–3507, 2003.

[HTWN10] W. P. M. H. Heemels, Andrew R. Teel, Nathan van de Wouw, and Dragan Nešić. Networked control systems with communication constraints: Tradeoffs between transmission intervals, delays and performance. *IEEE Transactions on Automatic Control*, 55(8):1781–1796, 2010.

[HVL05] Dimitrios Hristu-Varsakelis and William S. Levine, editors. *Handbook of Networked and Embedded Control Systems*. Birkhäuser, Boston, MA, 2005.

[HWG+10] W. P. M. H. Heemels, Nathan van de Wouw, Rob H. Gielen, M. C. F. Donkers, Laurentiu Hetel, Sorin Olaru, Mircea Lazar, Jamal Daafouz, and Silviu Niculescu. Comparison of overapproximation methods for stability analysis of networked control systems. In *Proceedings of the 13th International Conference on Hybrid Systems: Computation and Control*, pages 181–190, 2010.

[IGL10] Michal Izák, Daniel Görges, and Steven Liu. Stabilization of systems with variable and uncertain sampling period and time delay. *Nonlinear Analysis: Hybrid Systems*, 4(2):291–305, 2010.

[JÖ6] Ulf T. Jönsson. A lecture on the S-procedure. *Lecture Note at the Royal Institute of technology, Sweden*, pages 1–20, 2006.

[JHL+12] A. Jokic, R. M. Hermans, M. Lazar, A. Alessio, P. P. J. van den Bosch, I. A. Hiskens, and A. Bemporad. Assessment of non-centralized model predictive control techniques for electrical power networks. *International Journal of Control*, 85(8):1162–1177, 2012.

[JJ10] Mikael Johansson and Riku Jäntti. *Wireless Networking for Control: Technologies and Models*, chapter 2, pages 31–74. Springer-Verlag, Berlin Heidelberg, 2010.

[JTN05] Karl Henrik Johansson, Martin Törngren, and Lars Nielsen. *Vehicle Applications of Controller Area Network*, chapter 6, pages 741–765. Birkhäuser Boston, 2005.

[KBM96] Mayuresh V. Kothare, Venkataramanan Balakrishnan, and Manfred Morari. Robust constrained model predictive control using linear matrix inequalities. *Automatica*, 32(10):1361–1379, 1996.

[KS72] Huibert Kwakernaak and Raphael Sivan. *Linear Optimal Control Systems*. Wiley-Interscience, New York, NY, 1972.

[LA09] Hai Lin and Panos J. Antsaklis. Stability and stabilizability of switched linear systems: A survey of recent results. *IEEE Transactions on Automatic Control*, 54(2):308–322, 2009.

[LB02] Bo Lincoln and Bo Bernhardsson. LQR optimization of linear system switching. *IEEE Transactions on Automatic Control*, 47(10):1701–1705, 2002.

[LDWH12] S. J. L. M. van Loon, M. C. F. Donkers, N. van de Wouw, and W. P. M. H. Heemels. Stability analysis of networked control systems with periodic protocols and uniform quantizers. In *Proceedings of the 4th IFAC Conference on Analysis and Design of Hybrid Systems*, pages 186–191, 2012.

[LDWH13] S. J. L. M. van Loon, M. C. F. Donkers, N. van de Wouw, and W. P. M. H. Heemels. Stability analysis of networked and quantized linear control systems. *Nonlinear Analysis: Hybrid Systems*, 10:111–125, 2013.

[Lei09] Paulo Leitão. Agent-based distributed manufacturing control: A state-of-the-art survey. *Engineering Applications of Artificial Intelligence*, 22(7):979–991, 2009.

[Löf03] Johan Löfberg. *Minimax approaches to robust model predictive control*. PhD thesis, Department of Electrical Engineering, Linköping University, Sweden, 2003.

[Löf04] Johan Löfberg. YALMIP : A toolbox for modeling and optimization in MATLAB. In *Proceedings of the IEEE International Symposium on Computer Aided Control Systems Design*, pages 284–289, 2004.

[LFH12] Kun Liu, Emilia Fridman, and Laurentiu Hetel. Network-based control via a novel analysis of hybrid systems with time-varying delays. In *Proceedings of the 51st IEEE Conference on Decision and Control*, pages 3886–3891, 2012.

[LGOH09] Junqi Liu, Azwirman Gusrialdi, Dragan Obradovic, and Sandra Hirche. Study on the effect of time delay on the performance of distributed power grids with networked cooperative control. In *Proceedings of the 1st IFAC Workshop on Estimation and Control of Networked Systems*, pages 168–173, 2009.

[LH13] Dan Liu and Fei Hao. Decentralized event-triggered control strategy in distributed networked systems with delays. *International Journal of Control, Automation, and Systems*, 11(1):33–40, 2013.

[LHB09] Stefano Longo, Guldo Herrmann, and Phil Barber. Optimization approaches for controller and schedule codesign in networked control. In *Proceedings of the 6th IFAC Symposium on Robust Control Design*, pages 301–306, 2009.

[LHWB06] Mircea Lazar, W. P. M. H. Heemels, S. Weiland, and A. Bemporad. Stabilizing model predictive control of hybrid systems. *IEEE Transactions on Automatic Control*, 51(11):1813–1818, 2006.

[Lib03a] Daniel Liberzon. Hybrid feedback stabilization of systems with quantized signals. *Automatica*, 39:1543–1554, 2003.

[Lib03b] Daniel Liberzon. *Switching in Systems and Control.* Birkhäuser, Boston, 2003.

[LL10] Jan Lunze and Daniel Lehmann. A state-feedback approach to event-based control. *Automatica*, 46:211–215, 2010.

[LL11a] Daniel Lehmann and Jan Lunze. Event-based control with communication delays. In *Proceedings of the 18th IFAC World Congress*, pages 3262–3267, 2011.

[LL11b] Daniel Lehmann and Jan Lunze. Event-based output-feedback control. In *Proceedings of the 19th Mediterranean Conference on Control and Automation*, pages 982–987, 2011.

[LLJ12] Daniel Lehmann, Jan Lunze, and Karl H. Johansson. Comparison between sampled-data control, deadband control and model-based event-triggered control. In *Proceedings of the 4th IFAC Conference on Analysis and Design of Hybrid Systems*, pages 7–12, 2012.

[LMT01] Feng-Li Lian, James R. Moyne, and Dawn M. Tilbury. Performance evaluation of control networks: Ethernet, ControlNet, and DeviceNet. *IEEE Control Systems Magazine*, 21(1):66–83, 2001.

[LR06] Bo Lincoln and Anders Rantzer. Relaxing dynamic programming. *IEEE Transactions on Automatic Control*, 51(8):1249–1260, 2006.

[Lun14] Jan Lunze, editor. *Control Theory of Digitally Networked Dynamic Systems.* Springer, 2014.

[LVM07] Camilo Lozoya, Manel Velasco, and Pau Martí Colom. A 10-year taxonomy on prior work on sampling period selection for resource-constrained real-time control systems. In *Proceedings Work-in-Progress Session of the 19th Euromicro Conference on Real-Time Systems*, pages 29–32, 2007.

[LWHZ14] Qinyuan Liu, Zidong Wang, Xiao He, and D. H. Zhou. A survey of event-based strategies on control and estimation. *Systems Science & Control Engineering: An Open Access Journal*, 2(1):90–97, 2014.

[LZFA05] Hai Lin, Guisheng Zhai, Lei Fang, and Panos J. Antsaklis. Stability and $\mathcal{H}_\infty$ performance preserving scheduling policy for networked control systems. In *Proceedings of the 16th IFAC World Congress*, 2005.

[LZJ09] Yuan Li, Qingling Zhang, and Chong Jing. Stochastic stability of networked control systems with time-varying sampling periods. *International Journal of Information and Systems Sciences*, 5(3-4):494–502, 2009.

[LZNS10] S. Leirens, C. Zamora, R. R. Negenborn, and B. De Schutter. Coordination in urban water supply networks using distributed model predictive control. In *Proceedings of the American Control Conference*, pages 3957–3962, 2010.

[MA03] Luis A. Montestruque and Panos J. Antsaklis. Stochastic stability for model-based networked control systems. In *Proceedings of the American Control Conference*, pages 4119–4124, 2003.

[MC08] Huijun Gao Xiangyu Meng and Tongwen Chen. Stabilization of networked control systems with a new delay characterization. *IEEE Transactions on Automatic Control*, 53(9):2142–2148, 2008.

[MC14] Xiangyu Meng and Tongwen Chen. Event detection and control co-design of sampled-data systems. *International Journal of Control*, 87(4):777–786, 2014.

[Mey00] Carl Dean Meyer. *Matrix Analysis and Applied Linear Algebra*. Society for Industrial and Applied Mathematics (SIAM), Philadelphia, PA, 2000.

[MH09] Adam Molin and Sandra Hirche. On LQG joint optimal scheduling and control under communication constraints. In *Proceedings of the 48th IEEE Conference on Decision and Control and 28th Chinese Control Conference*, pages 5832–5838, 2009.

[MH10] Adam Molin and Sandra Hirche. Suboptimal event based control of linear systems over lossy channels. In *Proceedings of the 2nd IFAC Workshop on Estimation and Control of Networked Systems*, pages 55–60, 2010.

[MH11] Adam Molin and Sandra Hirche. Optimal design of decentralized event-triggered controllers for large-scale systems with contention-based communication. In *Proceedings of the 50th IEEE Conference on Decision and Control and European Control Conference*, pages 4710–4716, 2011.

[MH12] Adam Molin and Sandra Hirche. Adaptive event-triggered control over a shared network. In *Proceedings of the 51st IEEE Conference on Decision and Control*, pages 6591–6596, 2012.

[Mir07] Leonid Mirkin. Some remarks on the use of time-varying delay to model sample-and-hold circuits. *IEEE Transactions on Automatic Control*, 52(6):1109–1112, 2007.

[MR03] David Q. Mayne and S. Raković. Model predictive control of constrained piecewise affine discrete-time systems. *International Journal of Robust and Nonlinear Control*, 13(3-4):261–279, 2003.

[MR09] Karl Mårtensson and Anders Rantzer. Gradient methods for iterative distributed control synthesis. In *Proceedings of the 48th IEEE Conference on Decision and Control and the 28th Chinese Control Conference*, pages 549–554, 2009.

[MRRS00] David Q. Mayne, J.B. Rawlings, C.V. Rao, and P.O.M. Scokaert. Constrained model predictive control: Stability and optimality. *Automatica*, 36(6):789–814, 2000.

[MT07] James R. Moyne and Dawn M. Tilbury. The emergence of industrial control networks for manufacturing control, diagnostics, and safety data. *Proceedings of the IEEE*, 95(1):29–47, 2007.

[NBW98] Johan Nilsson, Bo Bernhardsson, and Björn Wittenmark. Stochastic analysis and control of real-time systems with random time delays. *Automatica*, 34(1):57–64, 1998.

[NE03] Girish N. Nair and Robin J. Evans. Exponential stabilisability of finite-dimensional linear systems with limited data rates. *Automatica*, 39(4):585–593, 2003.

[NH09] Payam Naghshtabrizi and João P. Hespanha. Analysis of distributed control systems with shared communication and computation resources. In *Proceedings of the 2009 American Control Conference*, pages 3384–3389, 2009.

[NHB05] Thomas Nolte, Hans Hansson, and Lucia Lo Bello. Automotive communications – past, current and future. In *Proceedings of the 10th IEEE Conference on Emerging Technologies and Factory Automation*, pages 985–992, 2005.

[NHT07] Payam Naghshtabrizi, João P. Hespanha, and Andrew R. Teel. Stability of delay impulsive systems with application to networked control systems. In *Proceedings of the 2007 American Control Conference*, pages 4899–4904, 2007.

[NHT08] Payam Naghshtabrizi, João P. Hespanha, and Andrew R. Teel. Exponential stability of impulsive systems with application to uncertain sampled-data systems. *System & Control Letters*, 57(5):378–385, 2008.

[NJWY09] Yugang Niu, Tinggang Jia, Xingyu Wang, and Fuwen Yang. Output-feedback control design for NCSs subject to quantization and dropout. *In-*

formation Sciences, 179(21):3804–3813, 2009.

[NL09] Dragan Nešić and Daniel Liberzon. A unified framework for design and analysis of networked and quantized control systems. *IEEE Transactions on Automatic Control*, 54(4):732–747, 2009.

[NP97] Vesna Nevistić and James A. Primbs. Finite receding horizon linear quadratic control: A unifying theory for stability and performance analysis. Technical Report CIT-CDS 97-001, California Institute of Technology, Control and Dynamical Systems, Pasadena, CA, USA, 1997.

[NSH08] Rudy R. Negenborn, Bart De Schutter, and J. Hellendoorn. Multi-agent model predictive control for transportation networks: Serial versus parallel schemes. *Engineering Applications of Artificial Intelligence*, 21(3):353–366, 2008.

[NT04a] Dragan Nešić and A. R. Teel. Input-output stability properties of networked control systems. *IEEE Transactions on Automatic Control*, 49(10):1650–1667, 2004.

[NT04b] Dragan Nešić and A. R. Teel. Input-to-state stability of networked control systems. *Automatica*, 40(12):2121–2128, 2004.

[OFL04] Petter Ögren, Edward Fiorelli, and Naomi Ehrich Leonard. Cooperative control of mobile sensor networks: Adaptive gradient climbing in a distributed environment. *IEEE Transactions on Automatic Control*, 49(8):1292–1302, 2004.

[PMFJ12] Pangun Park, Piergiuseppe Di Marco, Carlo Fischione, and Karl Henrik Johansson. Delay distribution analysis of wireless personal area networks. In *Proceedings of the 51st IEEE Conference on Decision and Control*, pages 5864–5869, 2012.

[PN00] James A. Primbs and Vesna Nevistić. Feasibility and stability of constrained finite receding horizon control. *Automatica*, 36(7):965–971, 2000.

[PSM04] Bert Pluymers, Johan Suykens, and Bart De Moor. Robust finite-horizon MPC using optimal worst-case closed-loop predictions. In *Proceedings of the 43rd IEEE Conference on Decision and Control*, pages 2503–2508, 2004.

[PSM05] Bert Pluymers, Johan A. K. Suykens, and Bart De Moor. Min-max feedback MPC using a time-varying terminal constraint set and comments on "efficient robust constrained model predictive control with a time-varying terminal constraint set". *Systems & Control Letters*, 54(12):1143–1148, 2005.

[PTNA11] Romain Postoyan, Paulo Tabuada, Dragan Nešić, and Adolfo Anta. Event-triggered and self-triggered stabilization of distributed networked control systems. In *Proceedings of the 50th IEEE Conference on Decision and Con-*

trol and European Control Conference, pages 2565–2570, 2011.

[PY13] Chen Peng and Tai Cheng Yang. Event-triggered communication and H_∞ control co-design for networked control systems. *Automatica*, 49(5):1326–1332, 2013.

[RAL13a] Sven Reimann, Sanad Al-Areqi, and Steven Liu. An event-based online scheduling approach for networked embedded control systems. In *Proceedings of the American Control Conference*, pages 5326–5331, 2013.

[RAL13b] Sven Reimann, Sanad Al-Areqi, and Steven Liu. Output-based control and scheduling codesign for control systems sharing a limited resource. In *Proceedings of the 52nd IEEE Conference on Decision and Control*, pages 4042–4047, 2013.

[Ran06] Anders Rantzer. Relaxed dynamic programming in switching systems. *IEE Proceedings – Control Theory & Applications*, 153(5):567–574, 2006.

[Ric03] Jean-Pierre Richard. Time-delay systems: An overview of some recent advances and open problems. *Automatica*, 39(10):1667–1694, 2003.

[RJ09] Maben Rabi and Karl H. Johansson. Scheduling packets for event-triggered control. In *Proceedings of the European Control Conference*, pages 3779–3784, 2009.

[RS00] Henrik Rehbinder and Martin Sanfridson. Integration of off-line scheduling and optimal control. In *Proceedings of the 12th Euromicro Conference on Real-Time Systems*, pages 137–143, 2000.

[RS04] Henrik Rehbinder and Martin Sanfridson. Scheduling of a limited communication channel for optimal control. *Automatica*, 40(3):491–500, 2004.

[RSBJ11] Chithrupa Ramesh, Henrik Sandberg, Lei Bao, and Karl Henrik Johansson. On the dual effect in state-based scheduling of networked control systems. In *Proceedings of the 2011 American Control Conference*, pages 2216–2221, 2011.

[RSJ13] Chithrupa Ramesh, Henrik Sandberg, and Karl H. Johansson. Design of state-based schedulers for a network of control loops. *IEEE Transactions on Automatic Control*, 58(8):1962–1975, 2013.

[RVAL15] Sven Reimann, Duc Hai Van, Sanad Al-Areqi, and Steven Liu. Stability analysis and PI control synthesis under event-triggered communication. In *Proceedings of the 2015 European Control Conference*, pages 2179–2184, 2015.

[San04] Martin Sanfridson. *Quality of control and real-time scheduling*. PhD thesis, Royal Institute of Technology, Department of Machine Design, Stockholm, Sweden, 2004.

[SB09] Joëlle Skaf and Stephen Boyd. Analysis and synthesis of state-feedback controllers with timing jitter. *IEEE Transactions on Automatic Control*, 54(3):652–657, 2009.

[SB11] Li Shanbin and Xu Bugong. Co-design of event generator and controller for event-triggered control system. In *Proceedings of the 30th Chinese Control Conference*, pages 175–179, 2011.

[SBK05] Bharath Sundararaman, Ugo Buy, and Ajay D. Kshemkalyani. Clock synchronization for wireless sensor networks: A survey. *Ad Hoc Networks*, 3(3):281–323, 2005.

[SCEP09] Soheil Samii, Anton Cervin, Petru Eles, and Zebo Peng. Integrated scheduling and synthesis of control applications on distributed embedded systems. In *Proceedings of Design, Automation & Test in Europe (DATE)*, pages 57–62, 2009.

[SDHW10] J. J. C. van Schendel, M. C. F. Donkers, W. P. M. H. Heemels, and N. van de Wouw. On dropout modelling for stability analysis of networked control systems. In *Proceedings of the American Control Conference*, pages 555–561, 2010.

[SG05a] Zhendong Sun and S. S. Ge. Analysis and synthesis of switched linear control systems. *Automatica*, 41(2):181–195, 2005.

[SG05b] Zhendong Sun and Shuzhi Sam Ge. *Switched Linear Systems: Control and Design.* Communications and Control Engineering. Springer, 2005.

[SL13] Christian Stöcker and Jan Lunze. Distributed event-based control of physically interconnected systems. In *Proceedings of the 52nd IEEE Conference on Decision and Control*, pages 7376–7383, 2013.

[SP05] Sigurd Skogestad and Ian Postlethwaite. *Multivariable Feedback Control: Analysis and Design.* John Wiley & Sons, Chichester, England, 2005.

[SQ10] Y. Sun and S. Qin. Stability of networked control systems with packet dropout: an average dwell time approach. *IET Control Theory and Applications*, 5(1):47–53, 2010.

[SR98] Pierre O. M. Scokaert and James B. Rawlings. Constrained linear quadratic regulation. *IEEE Transactions on Automatic Control*, 43(8):1163–1169, 1998.

[SSF+07] Luca Schenato, Bruno Sinopoli, Massimo Franceschetti, Kameshwar Poolla, and S. Shankar Sastry. Foundations of control and estimation over lossy networks. *Proceedings of the IEEE*, 95(1):163–187, 2007.

[Suh08] Young Soo Suh. Stability and stabilization of nonuniform sampling systems. *Automatica*, 44(12):3222–3226, 2008.

[SY09] Yang Shi and Bo Yu. Output feedback stabilization of networked control systems with random delays modeled by Markov chains. *IEEE Transactions on Automatic Control*, 54(7):1668–1674, 2009.

[Tab07] Paulo Tabuada. Event-triggered real-time scheduling of stabilizing control tasks. *IEEE Transactions on Automatic Control*, 52(9):1680–1685, 2007.

[TN08] Mohammad Tabbara and Dragan Nešić. Input-output stability of networked control systems with stochastic protocols and channels. *IEEE Transactions on Automatic Control*, 53(5):1160–1175, 2008.

[TW11] Andrew S. Tanenbaum and David J. Wetherall. *Computer Networks*. Pearson, Boston, 5th edition, 2011.

[VD05] Vladimeros Vladimerou and Geir Dullerud. *Wireless Control with Bluetooth*, chapter 6, pages 779–792. Birkhäuser, Boston, 2005.

[WARL15] Benjamin Watkins, Sanad Al-Areqi, Sven Reimann, and Steven Liu. Event-based control of constrained discrete-time linear systems with guaranteed performance. *International Journal of Sensors, Wireless Communications, and Control*, 5(2):72–80, 2015.

[WC07] Jing Wu and Tongwen Chen. Design of networked control systems with packet dropouts. *IEEE Transactions on Automatic Control*, 52(7):1314–1319, 2007.

[Wit66] Hans S. Witsenhausen. A class of hybrid-state continuous-time dynamic systems. *IEEE Transactions on Automatic Control*, AC-11(2):161–167, 1966.

[WK03] Zhaoyang Wan and Mayuresh V. Kothare. An efficient off-line formulation of robust model predictive control using linear matrix inequalities. *Automatica*, 39(5):837–846, 2003.

[WL11] Xiaofeng Wang and Michael D. Lemmon. Event-triggering in distributed networked control systems. *IEEE Transactions on Automatic Control*, 56(3):586–601, 2011.

[WNH12] Nathan van de Wouw, Dragan Nešić, and W. P. M. H. Heemels. A discrete-time framework for stability analysis of nonlinear networked control systems. *Automatica*, 48(6):1144–1153, 2012.

[WS12] Zhi-Wen Wang and Hong-tao Sun. Control and scheduling co-design of networked control system: Overview and directions. In *Proceedings of the International Conference on Machine Learning and Cybernetics*, pages 816–824, 2012.

[WY01] Gregory C. Walsh and Hong Ye. Scheduling of networked control systems. *IEEE Control Systems Magazine*, 21(1):57–65, 2001.

[WYB02] Gregory C. Walsh, Hong Ye, and L. G. Bushnell. Stability analysis of networked control systems. *IEEE Transactions on Control Systems Technology*, 10(3):438–446, 2002.

[XH04] Yonggang Xu and João P. Hespanha. Optimal communication logics in networked control systems. In *Proceedings of the 43rd IEEE Conference on Decision and Control*, pages 3527–3532, 2004.

[XJ08] Zhang Xiang and Xiao Jian. Communication and control co-design for networked control system in optimal control. In *Proceedings of the 12th WSEAS International Conference on Systems*, pages 698–703, 2008.

[XS06] Feng Xia and Youxian Sun. Control-scheduling codesign: A perspective on integrating control and computing. *Dynamics of Continuous, Discrete and Impulsive Systems*, 13(S1):1352–1358, 2006.

[XS08] Feng Xia and Youxian Sun. *Control and Scheduling Codesign: Flexible Resource Management in Real-Time Control Systems.* Zhejiang University Press, Hangzhou and Springer -Verlag GmbH Berlin Heidelberg, 2008.

[YA13] Han Yu and Panos J. Antsaklis. Event-triggered output feedback control for networked control systems using passivity: Achieving $\mathcal{L}_2$ stability in the presence of communication delays and signal quantization. *Automatica*, 49(1):30 – 38, 2013.

[Yan06] Tai C. Yang. Networked control system: A brief survey. *IEE Proceedings – Control Theory & Applications*, 153(4):403–412, 2006.

[YHP04] Dong Yue, Qing-Long Han, and Chen Peng. State feedback controller design of networked control systems. *IEEE Transactions on Circuits and Systems*, 51(11):640–644, 2004.

[YSLG11] Rongni Yang, Peng Shi, Guo-Ping Liu, and Huijun Gao. Networked-based feedback control for systems with mixed delays based on quantization and dropout compensation. *Automatica*, 47(12):2805–2809, 2011.

[YWCH04] Mei Yu, Long Wang, Tianguang Chu, and Fei Hao. An LMI approach to networked control systems with data packet dropout and transmission delays. In *Proceedings of the 43rd IEEE Conference on Decision and Control*, pages 3545–3550, 2004.

[YWCH05] Mei Yu, Long Wang, Tianguang Chu, and Fei Hao. Stabilization of networked control systems with data packet dropout and transmission delays: Continuous time case. *European Journal of Control*, 11(1):40–49, 2005.

[YWL+11] Hehua Yan, Jiafu Wan, Di Li, Yuqing Tu, and Ping Zhang. Codesign of networked control systems: A review from different perspectives. In *Proceedings of the IEEE International Conference on Cyber Technology in Automation,*

Control, and Intelligent Systems, pages 84–90, 2011.

[Ż03] Stanislaw H. Żak. *Systems and Control.* Oxford University Press, NY, 2003.

[Zam08] S. Zampieri. Trends in networked control systems. In *Proceedings of the 17th IFAC World Congress*, pages 2886–2894, 2008.

[ZGK13] Lixian Zhang, Huijun Gao, and Okyay Kaynak. Network-induced constraints in networked control systems: A survey. *IEEE Transactions on Industrial Informatics*, 9(1):403–416, 2013.

[ZHV06] Lei Zhang and Dimitrios Hristu-Varsakelis. Communication and control co-design for networked control systems. *Automatica*, 42(6):953–958, 2006.

[ZLR08] Y. B. Zhao, G. P. Liu, and D. Rees. Integrated predictive control and scheduling co-design for networked control systems. *IET Control Theory Applications*, 2(1):07–15, 2008.

[ZSCH05] Liqian Zhang, Yang Shi, Tongwen Chen, and Biao Huang. A new method for stabilization of networked control systems with random delays. *IEEE Transactions on Automatic Control*, 50(8):1177–1181, 2005.

Zusammenfassung

Untersuchung zu robusten Codesign Methoden für vernetzte Regelungssysteme

In dieser Arbeit wird das Problem der Gestaltung eines Reglers und Schedulers für vernetzte Regelungssysteme diskutiert. Vernetzte Regelungssysteme, welche aus mehreren Regelstrecken und einem gemeinsamen Kommunikationsnetz mit unsicheren, zeitlich variierenden Übertragungszeiten bestehen, werden als geschaltete, polytopische Systeme mit beschränkter, additiver Unsicherheit modelliert. Das Scheduling wird im vorliegenden Modell als Umschalter modelliert. Hierbei werden mögliche nichtkonvexe Mengen aufgrund der unsicheren Übertragungszeiten durch Polytope überschätzt. Basierend auf dem resultierenden Modell des vernetzten Regelungssystems sowie einem zustandsbasierten Regler, wird das Problem des gemeinsamen Entwurfs der Schaltfolge $j(0), \ldots, j(\infty)$ und die Rückführungsfolge $\boldsymbol{K}(0), \ldots, \boldsymbol{K}(\infty)$ als ein Optimierungsproblem formuliert. Ziel der Optimierung ist die Minimierung des Worst-Case Wertes einer quadratischen Kostenfunktion mit unendlichem Zeithorizont. Zur Lösung dieses Optimierungsproblems werden fünf robuste Codesign-Methoden vorgeschlagen. Die Eigenschaften jeder Methode werden durch Simulation und experimentelle Fallstudien ausgewertet und in Bezug auf deren Regelgüte verglichen. Die Ergebnisse der Arbeit zeigen sowohl eine Verbesserung der Regelgüte als auch eine effizientere Nutzung der vorhandenen Ressourcen. Die vorgeschlagenen Methoden lassen sich wie folgt zusammenfassen:

Periodische Regelung und Scheduling. Um das Optimierungsproblem zu lösen, werden die Schaltfolge $j(0), \ldots, j(\infty)$ und die Rückführungsfolge $\boldsymbol{K}(0), \ldots, \boldsymbol{K}(\infty)$ als p-periodische Folgen angenommen, also $j(k+p) = j(k)$ und $\boldsymbol{K}(k+p) = \boldsymbol{K}(k)$. Unter dieser Annahme wird das Codesign-Problem in das Problem der Bestimmung einer optimalen p-periodisch Schalt-Rückführ Sequenz σ^* überführt. Dies geschieht in zwei Schritten. Zunächst wird die Menge $\mathbb{S}_p$ aller zulässigen p-periodisch schaltenden Rückführsequenzen σ bestimmt. Als zulässig werden jene Sequenzen bezeichnet, welche das geschaltete System mit polytopischer Unsicherheit stabilisieren können. Da der Schalt-Index $j(k)$ zu einer endlichen Menge $\mathbb{M}$ gehört, werden alle p-periodischen Schaltsequenzen durch wiederholte Permutationen bestimmt. Die entsprechenden p periodischen Rückführungsmatrizen $\boldsymbol{K}(k)$ werden anhand eines LMI-Optimierungsproblems unter Verwendung p-periodischer Lyapunov Funktionen ermittelt. Nach Ermittlung der Menge $\mathbb{S}_p$ beginnt der zweite Schritt. Hierbei wird jedes Element der der Menge $\mathbb{S}_p$ in Bezug auf das Vorhandensein einer optimalen Lösung überprüft (ausführliche Suche). Um die vorausgesetzte Periodizität zu unterdrücken und somit zu einer Verbesserung der resultierenden Regelgüte zu gelangen, wird die ausführliche Suche zu jedem Zeitpunkt t_k für den aktuellen Zustand $\boldsymbol{x}(k)$ durchgeführt. Zur Reduzierung des online-Berechnungsaufwandes der ausführlichen Suche, werden suboptimale Lösungen zugelassen. Dabei werden zwei relaxierte Versionen der periodischen Codesign-Strategie vorgeschlagen. Die Wirksamkeit dieser Strategien und der Einfluss der gewählten Periode p auf die resultierende Regelgüte werden durch Simulation für ein illustratives Beispiel sowie experimentell für eine Fallstudie ausgewertet.

Receding-Horizon Regelung und Scheduling. Anstatt Periodizität für die komplette Schalt- und Rückführungsfolge vorauszusetzen, wird die Kostenfunktion mit unendlichem Zeithorizont in die beiden Teile J_1 und J_2 zerlegt. Im zweiten Teil J_2 wird eine stabilisierende p-periodisch Schalt-Rückführ Sequenz σ angesetzt. Mit dieser Annahme wird aus der Kostenfunktion mit ursprünglich unendlichem Zeithorizont eine Kostenfunktion J_N mit dem endlichen Zeithorizont N. Somit lässt sich das Codesign-Problem in einem bewegten Zeithorizont durch Verwendung dynamischer Programmierung, unter Berücksichtigung des aktuellen Zustands $\boldsymbol{x}(k)$, lösen. Ein großer Teil der Berechnungen der dynamischen Programmierung kann offline erfolgen, allerdings wächst sowohl im online- als auch im offline Modus die Komplexität exponentiell mit dem Zeithorizont N. Praktisch gesehen ist die Optimierungsaufgabe dadurch unlösbar. Aus diesem Grund wird eine relaxierte dynamische Programmierung eingeführt. Folglich wird die Komplexität verringert, während sich die Regelgüte verschlechtert. Die Wirksamkeit der vorgeschlagenen Strategie und der Einfluss des Zeithorizonts N auf die resultierende Regelgüte wird sowohl simulativ für ein illustratives Beispiel als auch experimentell für eine Fallstudie ausgewertet.

Implementation-Aware Regelung und Scheduling. Bei dieser Strategie wird weder Periodizität der Schaltfolge vorausgesetzt noch eine Partitionierung der Kostenfunktion vorgenommen. Stattdessen wird ein zustandsbasiertes Schaltgesetz mit quadratischer Struktur eingeführt. Anhand des Schaltgesetzes wird das Codesign-Problem als LMI-Optimierungsproblem unter Verwendung der Lyapunov-Metzler Funktion formuliert. Der optimale Schalt-Index $j^*(k)$ und die optimale Rückführungsmatrix $\boldsymbol{K}^*(k)$ werden in jedem Zeitpunkt t_k durch Lösung des LMI Optimierungsproblems für den aktuellen Zustand $\boldsymbol{x}(k)$ ermittelt. Im Gegensatz zu den obengenannten Strategien, bei denen die Rückführungsmatrizen $\boldsymbol{K}(k)$ offline ohne Berücksichtigung des Zustandes $\boldsymbol{x}(k)$ berechnet werden, wird $\boldsymbol{K}(k)$ online entworfen. Der aktuelle Zustand $\boldsymbol{x}(k)$ wird explizit berücksichtigt. Da das vorgeschlagene Codesign-Problem online gelöst wird, muss die Lösbarkeit des Problems genauer analysiert werden und im Anschluss die Stabilität des geschlossenen Regelkreises bewiesen werden. Abschließend wird eine offline-Version der Strategie zur Verringerung des Rechenaufwands vorgeschlagen, welche wiederum mit einer Verschlechterung der Regelgüte einhergeht.

Ereignisbasierte Regelung und Scheduling. Die obengenannten Strategien erfordern die Messung aller Zustände von allen Regelstrecken zu jedem Zeitpunkt t_k zur Verarbeitung im Scheduler. Diese Voraussetzung kann einerseits in einigen Anwendungen nicht erfüllt werden und andererseits zu großen Scheduling-Aufwänden führen. Als Konsequenz daraus verschlechtert sich die resultierende Regelgüte. Als Alternative wird vorgeschlagen, für jede Regelstrecke einen Ereignisgenerator zur Entscheidung über die Sendung des aktuellen Zustands einzusetzen. In jedem Ereignisgenerator wird ein Schwellwert basiertes Ereignis-Auslösegesetz $\sigma_i(k)$, $\forall i \in \mathbb{M}$, implementiert. Der Schwellwert kann durch einen Entwurfsparameter $\lambda_i \in \mathbb{R}_0^+$ abgestimmt werden. Die Scheduling-Entscheidung basiert auf dem Resultat des zugeordneten Ereignis-Auslösegesetzes und nicht auf den Zustandsmessungen selbst. Durch die Einführung des Ereignis-Auslösegesetzes wird das Codesign-Problem in das Problem der Bestimmung von $\sigma_i(k)$ und der Rückführungsmatrizen $\boldsymbol{K}_i$ für alle $i \in \mathbb{M}$ überführt. Die beschriebene Strategie wird als ein LMI-Optimierungsproblem unter Verwendung des S-Verfahrens formuliert. Die Wirksamkeit der vorgeschlagenen Strategie und der Einfluss des Entwurfsparameters λ_i, $\forall i \in \mathbb{M}$ werden durch Simulation für ein illustratives Beispiel und experimentell für einen Fallstudie betrachtet.

Prediction-based Regelung und Scheduling. Bei den oben beschriebenen Codesign Methoden kann aufgrund der beschränkten Anzahl an verfügbaren Kommunikationskanälen zu jedem Zeitpunkt t_k nur eine Regelstrecke geregelt werden. Bei der nun vorgeschlagenen Strategie können trotz dieser Einschränkung alle Regelstrecken gleichzeitig geregelt werden. Zu diesem Zweck werden alle Regelungsaufgaben auf einem (und nicht auf M) eingebetteten Prozessor implementiert. Darüber hinaus wird ein modellbasierter Prädiktor zwecks Berechnung der Regelsignale auf dem eingebetteten Prozessor umgesetzt. Zur Bestimmung des Zustandsfehlers (Differenz zwischen aktuellem und die prädiziertem Zustand) werden die Zustände in den entsprechenden Ereignisgeneratoren ebenfalls ermittelt. Wenn der Zustandsfehler einen bestimmten Schwellwert überschreitet, wird die aktuelle Zustandsmessung der aktuell geregelten Regelstrecke auf den eingebetteten Prozessor zwecks Korrektur der prädizierten Zustände gesendet. Die Einführung modellbasierter Prädiktoren sowohl auf der Strecken- als auch auf der Regler-Seite führen zu einer weiteren Verbesserung der Regelgüte, was auch durch entsprechende Simulationsergebnisse bestätigt wird. Experimentelle Untersuchungen decken jedoch Nachteile bei der Regelung und dem Scheduling im Fall von Modellungenauigkeiten auf. In diesem Fall werden die Regelsignale anhand „falsch“ prädizierten Systemzuständen berechnet und auch mehr Ereignisse ausgelöst, als es erforderlich gewesen wäre.

Reflexion. In der vorliegenden Arbeit werden fünf Codesign Methoden für vernetzte Regelungssysteme vorgeschlagen. Bei den ersten drei Methoden wird das Scheduling zentral durchgeführt, während bei den letzten beiden Methoden eine dezentrale Arbeitsweise vorliegt. Dementsprechend hängt die Auswahl eines bestimmten Verfahrens stark von der konkreten Anwendung ab. Für Anwendungen, bei denen ein zentrales Scheduling mit geringem Aufwand implementierbar ist, kann entweder die Receding-Horizon Strategie oder die Implementation-Aware Strategie verwendet werden. Andernfalls kann die Ereignisbasierte Strategie (oder die Prediction-based Strategie im Fall eines genauen Modells der Systemdynamik) zum Einsatz kommen. Obwohl man bei den simulativen und experimentellen Ergebnissen nicht von einem Beweis sprechen kann, bieten sie einen sehr guten Überblick über die Eigenschaften einer bestimmten Strategie. Hierbei wird es deutlich, dass die vorgeschlagenen Strategien Vorteile gegenüber der weit verbreiteten TDMA-Strategie aufweisen. Gleiches gilt bezüglich anderer Strategien, die in der Literatur (z.B. die OPP-Strategie) behandelt werden. Ein interessantes Thema für zukünftige Arbeiten bietet die auf analytischem Wege festgelegte der Höhe der induzierten Suboptimalität.

Curriculum Vitae

Personal Data

Name	Sanad Al-Areqi
Born	01 January 1983 in Taiz, Yemen
Address	Pariser Straße 290 67663 Kaiserslautern, Germany
E-Mail	alareqi@eit.uni-kl.de

Education

10/2008 – 10/2010 **University of Kaiserslautern – Germany**

Area: Elektrtotechnik und Informationstechnik (Electrical and Computer Engineering)

Field: Automatisierungstechnik (Automation)

Degree: Master of Science (M. Sc.)

Thesis: Robust Control and Scheduling Codesign for Networked Embedded Control Systems

10/1999 – 07/2004 **University of Hadhramout – Yemen**

Area: Communications Engineering

Degree: Bachelor of Science (B. Sc.)

Thesis: Implementing Code-Division Multiple Access using Frequency-Hopping for Mobile Communication

Professional Experience

11/2010 – 08/2015 **University of Kaiserslautern – Germany**

Department of Electrical and Computer Engineering
Institute of Control Systems

Research Associate

10/2004 – 07/2008 **University of Hadhramout – Yemen**
Department of Communications Engineering
Instructor and Assistant Lecturer

Kaiserslautern, May 2015

Sanad Al-Areqi

In der Reihe „*Forschungsberichte aus dem Lehrstuhl für Regelungssysteme*“, herausgegeben von Steven Liu, sind bisher erschienen:

1	Daniel Zirkel	Flachheitsbasierter Entwurf von Mehrgrößenregelungen am Beispiel eines Brennstoffzellensystems ISBN 978-3-8325-2549-1, 2010, 159 S. 35.00 €
2	Martin Pieschel	Frequenzselektive Aktivfilterung von Stromoberschwingungen mit einer erweiterten modellbasierten Prädiktivregelung ISBN 978-3-8325-2765-5, 2010, 160 S. 35.00 €
3	Philipp Münch	Konzeption und Entwurf integrierter Regelungen für Modulare Multilevel Umrichter ISBN 978-3-8325-2903-1, 2011, 183 S. 44.00 €
4	Jens Kroneis	Model-based trajectory tracking control of a planar parallel robot with redundancies ISBN 978-3-8325-2919-2, 2011, 279 S. 39.50 €
5	Daniel Görges	Optimal Control of Switched Systems with Application to Networked Embedded Control Systems ISBN 978-3-8325-3096-9, 2012, 201 S. 36.50 €
6	Christoph Prothmann	Ein Beitrag zur Schädigungsmodellierung von Komponenten im Nutzfahrzeug zur proaktiven Wartung ISBN 978-3-8325-3212-3, 2012, 118 S. 33.50 €
7	Guido Flohr	A contribution to model-based fault diagnosis of electro-pneumatic shift actuators in commercial vehicles ISBN 978-3-8325-3330-0, 2013, 139 S. 34.00 €

8	Jianfei Wang	Thermal Modeling and Management of Multi-Core Processors ISBN 978-3-8325-3699-2, 2014, 144 S.	35.50 €
9	Stefan Simon	Objektorientierte Methoden zum automatisierten Entwurf von modellbasierten Diagnosesystemen ISBN 978-3-8325-3940-5, 2015, 197 S.	36.50 €
10	Sven Reimann	Output-Based Control and Scheduling of Resource-Constrained Processes ISBN 978-3-8325-3980-1, 2015, 145 S.	34.50 €
11	Tim Nagel	Diagnoseorientierte Modellierung und Analyse örtlich verteilter Systeme am Beispiel des pneumatischen Leitungssystems in Nutzfahrzeugen ISBN 978-3-8325-4157-6, 2015, 306 S.	49.50 €